GIS for business and service planning

GIS for business and service planning

Edited by

Paul Longley and Graham Clarke

JOHN WILEY & SONS, INC.

New York • Chichester • Weinheim • Brisbane • Singapore • Toronto

Library of Congress Cataloging-in-Publication Data:

A catalogue entry for this title is available from the Library of Congress

ISBN 0-470-23510-1

Printed in the United States of America

10 9 8 7 6 5 4 3 2

Contents

Preface

The idea of a book on geographical information systems (GIS) for business and service planning was first mooted at a two-day seminar held in Leeds in April 1993, which was jointly sponsored by the UK Economic and Social Research Council (ESRC) and the Association for Geographical Information (AGI). The purpose of this two-day seminar was twofold: first, to present the latest ideas on data sources, visualization and analysis for GIS users in all areas of business (retailing, direct marketing, utility management, finance, transport, etc.); and, secondly, to bring together academics carrying out GIS-based research with practitioners and consultants to share experiences, offer new ideas and help formulate a strategy for future research.

The seminar was held at an interesting time. First, the 1980s had seen a phenomenal growth of interest in GIS, both within the academic research base and (particularly in the late 1980s) in the wider economy, and yet it was possible to discern the first signs of disillusionment with the technology as successive waves of bland descriptive applications failed to deliver lasting insights into applied problems. Secondly, a range of data products arising from the 1991 UK Census was coming on stream, and academics and businesses alike were beginning to explore new ways of 'adding value' to raw geographically referenced scores in an applied yet consistent analysis. Thirdly, the seminar followed a four-and-a-half-year period in which the ESRC had been involved in promoting applied GIS research through its Regional Research Laboratory Initiative, the US had instituted its National Center for Geographic Information and Analysis, and other countries such as the Netherlands and Japan had devised major programmes to encourage an eminently applicable technology to actually become applied to 'real world' problems. Many questions emerged from the discussions. What practical use are GIS in business and service planning? How may GIS be customized to specific user requirements? How might geodemographic analysis be developed to tackle the applied research challenges of the 1990s? Is the GIS revolution of lasting significance for the development of applied spatial analysis?

The chapters in this book were all written after the seminar, in the light of the exchanges which followed the various position statements. They analyse the linkages between technological change, analytical innovation and information customization which are now beginning to stimulate the wider adoption of GIS as a management and applied research tool. The book is intended to be one of the first in what is probably the most rapidly developing and dynamic area of GIS, and seeks to evaluate the experiences and views of the acknowledged leading world experts in the field.

Paul Longley
Graham Clarke

Bristol and Leeds, August 1994

Acknowledgements

The editors would like to thank a number of people for their contributions to this book. First, and foremost, we would like to thank all the speakers and delegates for the 'GIS in Business' seminar held at the Holiday Inn, Leeds, in April 1993. The lively discussion generated at this meeting provided unambiguous evidence that the GIS community was ready to embrace business applications and opportunities. We are also very grateful to Vanessa Lawrence of GeoInformation International for commissioning the project, for advising us on gaps in our original intended coverage (along with a set of anonymous referees), and for showing continued enthusiasm for its subject matter. The original diagrams were drawn by Simon Godden in Bristol and Alison Manson in Leeds with their characteristic care and adherence to tight deadlines. Lisa Rivers, Liz Humphries and Catherine Mckenna typed many of the tables. Financially, we acknowledge the support of the Association of Geographic Information (AGI) and of the Economic and Social Research Council (ESRC) through its seminar grant H-501-26-5009.

The editors would additionally like to thank the following individuals and organizations, which kindly gave permission to reproduce figures, diagrams and plates: Ordnance Survey (Plate 2.1 - based on 1988 Ordnance Survey figures, 1:10 000 map and Address Point, with the permission of the Controller of Her Majesty's Stationery Office © Crown Copyright); MIT Press Ltd, Cambridge, Mass. (Figure 5.1); CACI Ltd (Figure 6.1: special thanks for comments on the accompanying text by Jill Collins, Manager, Public Services Group); CCN Marketing Ltd (Figure 6.3; Plate 6.1; Plate 6.2); Polk Direct Ltd (Figure 6.11; Table 6.9); Kogan Page Ltd (Figure 6.13; Figure 6.14; Figure 6.15); Selected graphic images for front cover courtesy of Environmental Systems Research Institute, Inc. Copyright © 1995 Environmental Systems Research Institute, Inc. All rights reserved; and GeoInfo Systems for the tables from which Appendix B is derived.

Paul Longley
Department of Geography
University of Bristol

Graham Clarke
School of Geography
University of Leeds

List of Contributors

Peter Batey Lever Professor of Town and Regional Planning, Department of Civic Design, University of Liverpool, PO Box 147, Liverpool L69 3BX.

Mark Birkin Research Director, GMAP Ltd and Lecturer in Geography, University of Leeds, Leeds LS2 9JT.

Peter Brown Senior Lecturer, Department of Civic Design, University of Liverpool, PO Box 147, Liverpool L69 3BX.

Graham Clarke Lecturer in Geography, School of Geography, University of Leeds, Leeds LS2 9JT.

Martin Clarke Managing Director, GMAP Ltd, University of Leeds, Cromer Terrace, Leeds LS2 9JU and Professor of Spatial Decision Support Systems, School of Geography, University of Leeds, Leeds LS2 9JT.

Paul Cresswell Director, Spa Marketing Systems Ltd, 1 Warwick Street, Leamington Spa, Warwickshire CV32 5LW.

Paul Longley Reader in Geography, Department of Geography, University of Bristol, University Road, Bristol BS8 1SS.

David Maguire Technical Director, ESRI (UK), 23 Woodford Road, Watford WD1 1PB.

David Martin Lecturer in Geography, Department of Geography, University of Southampton, Southampton SO17 1BJ.

Ian Masser Professor of Town and Regional Planning, Department of Town and Regional Planning, University of Sheffield, Western Bank, Sheffield S10 2TN.

Stan Openshaw Professor of Human Geography, School of Geography, University of Leeds, Leeds LS2 9JT.

Nora Sherwood Editor, *Business Geographics*, GIS World Inc., 155 East Boardwalk Drive, Suite 250, Fort Collins, CO 80525, USA.

Robin Waters Data Products Manager, GeoInformation International, 307 Cambridge Science Park, Milton Road, Cambridge CB4 4ZD.

Introduction

1

Applied geographical information systems: developments and prospects

Paul Longley and Graham Clarke

1.1 Introduction

GIS has had a short but impressive history within academic geography, and many of the models, techniques and data-enhancement methods that have been pioneered in geography and planning are now making the transition from *potential* applications (Masser and Blakemore 1991) to applied analysis. There are many good reviews of the development of GIS in terms of principles, management and applications (e.g. Martin 1991; Cassettari 1993; Maguire *et al.* 1991), including two concerning business applications (Castle 1993a; Grimshaw 1994). GIS is also itself big business, with the commercial GIS industry comprising a number of major vendors each vying for increased market share. But GIS has now reached a crossroads in terms of potential directions it may take in the future. Some authors are claiming that it has not had the impact that the GIS community promised in the 1980s (e.g. Allinson 1993), and that greater spatial (geographical) analysis functions are required to guide GIS into a new era of applications-led activity (Fotheringham and Rogerson 1994; Birkin *et al.* 1995).

These observations certainly seem to be appropriate in the context of business and service planning. Despite the involvement of business in organizations such as the UK Association of Geographic Information and the innovation in the United States and Europe of annual conferences on 'GIS in Business', this community has yet to take GIS on board to the same extent as development managers in environmental fields, utility managers and those working in some areas of local government. The situation is changing, however, and the aim of this book is to explore and promote the use of GIS in the area of business and service planning through both methodological contributions and evidence of applied benefits. In short, this book is about the relevance of GIS to business and service planning in an era in which spatial analysis is beginning to unlock the wider potential of GIS.

1.2 Business and technology

Businesses and services function in society through their rational use of resources in order to fulfil economic and social objectives. Since the 1980s, there has been greater recognition that such activities require careful *planning*, be this in relation to private sector management of budgets concerning industrial strategy, or the allocation of scarce public resources between competing service demands. In fact, as industrial societies become still more complex and service-based, so the traditional distinction between the private and the public sectors is becoming blurred, and there is some convergence of management objectives across ever wider areas of business and service activity. Planning activities have thus attained a higher profile since the 1980s as an increasingly important use of human resources, although this is not the only way in which the effectiveness of planning has been enhanced.

During this period, technological change has also had a profound effect upon all significant world economies, particularly through the continuing real and absolute costs of computer technology in general and computer processing power in particular. These changes have been accompanied by massive development of computer software industries, which have harnessed hardware developments to improved methods for manipulating information and have created 'user-friendly' ways in which the wider community of information users can access information. Planning requires information, since it is information that provides the context for management decision-making (Beaumont 1991). Together, hardware and software developments have fostered profound changes in the ways in which information may be input, stored, retrieved and analysed using powerful yet distributed database management systems (DBMSs). This has had profound implications for the way in which information is viewed as a corporate or institutional resource (Openshaw and Goddard 1987), and furthermore has focused attention on the ways in which manipulation and analysis of raw data can 'add value' for planning purposes. In short, information has become a tradable commodity and a strategic resource. Routine analysis of such information is carried out using spreadsheets, business graphics and even word processing packages, while more advanced and possibly specialist software tools include decision support and expert systems.

A further major change in many business and service organizations is the increasing evidence of a change in our regard to locational decision-making. It is often argued that the retail sector has long been convinced that geography is important – the often-quoted maxim being that the three most important factors in retail success are 'location, location – and location'. As consumer-dominated markets become more sophisticated and competitive, so opportunities and threats will increasingly have a spatial dimension. It is perhaps surprising that the innovation of business graphics has not brought with it a greater preparedness to visualize space. The tide is turning, however: business graphics are becoming increasingly geographical. It is therefore natural that this provides an important and fast-growing area of application of GIS technology.

Despite a late start, business and service organizations are beginning to look towards GIS. Like other DBMSs, GIS allows the input, storage, retrieval and analysis of information, but unlike with other DBMSs, use is made of the property that this information pertains to unique points on the surface of the earth. Figure 1.1 (adapted from Martin 1991) illustrates how we might think of a GIS as a DBMS with an additional, geographical (spatial) 'shell' which enhances our ability to generate realistic yet accessible models of the real world. That is to say, space is the fundamental organizing principle through which information is organized, and geography is of central importance in the way in which data are manipulated and analysed. The various contributors to this book will argue that GIS is one of the newest and most valuable tools to be used as part of the ongoing 'information revolution' in the quest to gain competitive advantage in consumer-led markets.

GIS is now finding uses in an ever-widening range of commercial contexts. In the remaining chapters, we will analyse the linkages between technological change, analytical innovation and information customization which are stimulating the wider adoption of GIS as a management and applied research tool. GIS is an important global industry (Grimshaw 1994 reports that the GIS market reached $2 billion in 1992), and our aim is to report on advances and developments of international significance. As GIS becomes established in a variety of market and near-market contexts, the rate and pace of development is largely determined by three considerations: first, our ability to structure and manipulate the rapidly multiplying data sources into useful information while paying due attention to the problems of matching and merging geographical entities; secondly, the development of computer software to handle different classes of geographical problems, again with due regard to the special needs of spatial data handling; and thirdly, the prospects for disseminating developments in spatial data handling, in order that the broadest range of (preferably non-specialist) personnel can participate in decision-making based upon enhanced spatial analysis.

1.3 Structure of the book

The expert contributors to this book have been chosen from a wide variety of research and commercial contexts, but share common concerns with the deployment of geographical thinking and analysis in the solution of practical problems. In fact, just as business and public service share common management needs, so the management of geographical information spans a range of activities in management, consultancy, research and teaching which defy compartmentalization within any of the traditional 'academic' and 'commercial' categories. This very diversity highlights some important creative tensions and paradoxes in the ways in which developments are taking place.

The first paradox is that the innovation of machine-readable databases has not led to the routine development of geographical databases conducive to geographical analysis. David Martin and Paul Longley discuss how this is in

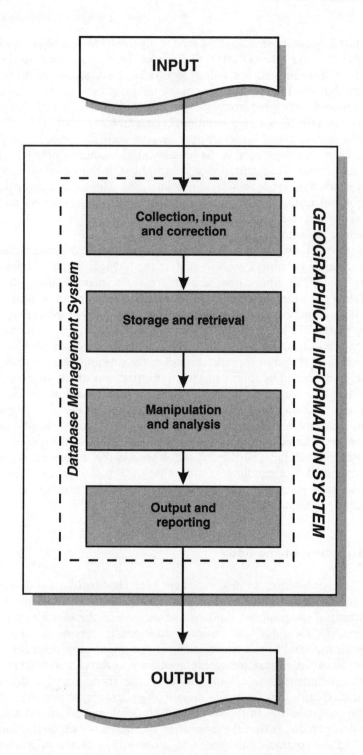

Fig. 1.1 The architecture of a GIS

part a consequence of the fragmentation of data holdings within large business and service organizations, and of the failure of data providers to consider the power and potential of explicit georeferencing. Robin Waters picks up these points in a trans-national context, in an assessment of public and private-domain data sources across the European Union and the commercial prospects for their supra-national integration. More fundamentally still, David Martin succinctly restates the problems that different and sometimes inconsistent areal units pose for geographical analysis, and suggests some new ways of depicting geographical information.

Secondly, the contributors illustrate the range of current perspectives on the analysis of spatial data. 'Business geographics' undoubtedly has its roots in the development of geodemographic indicators of customer purchasing behaviour. The development of geodemographics is described by Peter Batey and Peter Brown, who argue that the approach has significantly enhanced market analysis, through the provision of powerful discriminators of consumer behaviour. In its early forms, however, some would take issue with the dependence of the geodemographic approach upon surrogate information, which has sometimes been handled in a geographically dubious way. Stan Openshaw adopts a much more openly critical view of the achievements of applied geographical analysis in a contribution which argues that academic geography has been preoccupied with issues of analytical elegance at the expense of relevance and successful prediction. Mark Birkin takes a compromise position, suggesting that research and development using new techniques and methods hold the prospect of generating information and solutions which may be predictively successful whilst becoming more transparent and refined in theoretical terms. He also reviews significant new developments in geodemographics, particularly the development of 'Lifestyles' and psychographics.

Thirdly, tensions are apparent with regard to the software solutions that are advocated for geographical problems. The chapter by David Maguire is written from the standpoint of both a geographer and a major GIS software vendor and describes part of the range of geographical problems which may be tackled using proprietary software. Maguire describes a wide range of instances in which the purchase of a GIS by an organization can lead to the resolution of a range of geographical problems. The chapter by Paul Cresswell (again trained as a geographer) by contrast describes circumstances in which external consultancy and bureau services are necessary, either to solve problems directly, or to customize proprietary software to particular user needs. Graham Clarke and Martin Clarke adopt a more fundamental interpretation of 'customization', and describe a further range of circumstances in which the unique combination of problems and tasks requires the development of 'tailor-made' solutions by external consultancy services. The reconciliation of these different standpoints no doubt lies in part in a 'horses for courses' approach to different classes of geographical problem, but each of these contributions deserves close reading by those in business attempting to get to grips with the structure and linkage of geographical data for the first time.

The development of geographical analysis must be seen as the consequence of the interplay between methodology and applications development, and here it is clear that the organizational setting to business geographics is of critical importance, since it is this that fosters interaction between research and practice. Within the university sectors of most developed nations, GIS has come to be seen as one means of demonstrating the relevance and applicability of geographical ideas and of enhancing the status of geography itself. As part of this process, a number of initiatives have been developed in order to increase awareness of GIS outside geography, and to stimulate 'near-market' research using GIS methods emanating from the research base. A fourth creative difference concerns the different ways in which interaction between the GIS research base and actual and potential business and service users has been fostered. Nora Sherwood examines the extent of interaction in the US (see also Castle 1993a), and Graham Clarke *et al.* evaluate the recent history of 'near-market' GIS research with reference to the successful UK Regional Research Laboratory Initiative, the development of Teaching Company Schemes and the changing vertical and horizontal nature of research–industry links.

1.4 Applied GIS and geography

In sum, these various contributions amount to irrefutable evidence that the range of quantitative geographical techniques available in harness with GIS technology is now enhancing the status of geography in business and service planning, since geography demonstrably holds the key to further enhancement of management information systems. This conclusion is the more remarkable in that this activity is emanating from a discipline which has experienced significant and successive shifts in fashion over the last 20 years, shifts which have led to a progressive de-skilling of human geography in terms of ability to utilize quantitative geographical techniques and to build quantitative geographical models. In a critique of quantitative modelling in geography, Harvey (1989) describes its achievements as 'the proverbial hill of beans' (p. 213) which fail to deliver answers to important research questions. In part, the impingement of increased pressure upon universities in general and their geography departments has forced some reappraisal of the more 'ivory tower'-dominated views on the status and role of geography, and this can be no bad thing in so far as university geography departments have responsibilities to equip their graduates with a range of flexible skills in spatial data analysis for use in continually restructuring labour markets. More insidious, however, is the misinterpretation of comments such as Harvey's to imply that quantitative measurement is inherently 'invalid' and to refute the notion that valid domains for quantitative analysis actually exist. The range of issues and applications discussed here provides clear testimony that quantitative geography is most certainly *not* preoccupied with 'techniques that do not work to analyse problems that do not matter' (c.f. Chisholm 1971). A geography that does not show interest in real-world issues of popular concern is doomed to remain on the sidelines of academic respectability and perceived social relevance, and the various contributions to this book show how reinvigorated spatial analysis is central to the measurement and modelling of economic and social aspects of human behaviour.

Yet business geographics has as yet only a foothold in the marketplace, and geography has no strong practical arm in the way that economics has long been associated with business studies. The strong message here is nevertheless that quantitative geography will continue to develop the application and applicability of geographical thinking, and that the range of approaches described in this book will become increasingly central to the management and application of geographical data in business and service planning. In this sense this book is not just about the use of GIS in business and service planning, but is also about 'spreading the word' that geographers *do* have a vital role to play in providing techniques and analysis for strategic decision-making in a wide variety of public and private planning situations.

Part One

Population Data Sources: Measurement and Modelling

We began this book with the contention that the role of space in business and service planning has long been neglected, but that such neglect is untenable in the increasingly sophisticated customer-orientated markets of the 1990s. Recognition of the role of geography brings with it the problem that analysis of geographical (spatial) problems can be far from straightforward. Geographers have long recognized that 'spatial is special', and have partially understood the pitfalls of geographical data analysis; however, it is only in comparatively recent years, through the development first of computer and latterly of GIS technology, that it has proved possible to quantify some of these effects and to visualize their consequences. Geography is unique among sciences in not having 'natural' units of analysis (Openshaw 1984a): that is to say, for example, that there is not likely to be any exact and irrefutable measure of the distribution of potential customers for a new good or service, and thus any measure of market area is likely to be subjective. Businesses and service providers have been slow to grasp this concept, yet it is fundamental to the use of information for operational, tactical and strategic decision-making.

Illustration of this problem provides the starting point for David Martin and Paul Longley's discussion (Chapter 2) of how the areal units for which geographical data are collected and/or displayed can condition the results of any analysis. They describe different census, postal and *ad hoc* survey areas which may nevertheless be integrated within a GIS in consistent and meaningful ways. This chapter provides a foundation stone for much of the discussion of geodemographic data systems in Part Two of the book, with, for example, the integration of customer databases within the areal framework of census data providing one of the current challenges to spatial analysis. The very nature of competitive business lies in spotting niches and opportunities. Geographical niches are the more difficult to identify when market areas are blurred by the overlay of successive different geographies. Each geographical patterning of areal units may be thought of as a lens through which geographical patterning is viewed, and yet layering of different lenses on top of one another may do more to distort or obscure a composite geographical pattern than to clarify it. If correctly diagnosed, however, such problems are frequently tractable, and Martin and Longley develop an illustrative exposition of how postal, census and service area geographies might be reconciled within a consistent geographical framework. In practice, however, there may be additional problems in the integration of the often diverse data sources that are necessary for short- and longer-term decision-making, and these are explored using a 'real world' example. The technical and practical problems raised through this example are generic ones, and the reader is encouraged to relate these to other practical problems with which he/she has more immediate experience.

The need satisfactorily to resolve data issues in business applications of GIS provides a linking theme to Chapter 3 by Robin Waters. In the late twentieth century, consumer markets have become international or even global yet, frustratingly, data standardization and availability to inform international business strategies has lagged way behind the market. One of the most important international markets is that of the European Union (EU). Even a loosely federal Europe is still characterized by very considerable diversity of administrative structures, and this has unfortunate implications for the development of pan-European data standards and comparable areal units. Moreover, there is not yet a common framework for cost recovery or agreement as to how or whether third parties might be permitted to sell on 'value-added' databases. This is despite the best efforts of the European Commission's Eurostat, as described by Waters. Such data are nevertheless an important input to international business initiatives, and Waters provides a review of EU data sources and the potential and prospects for their integration. He also alludes to the development of pan-European 'geodemographic indicators', which potentially at least offer the prospect of enhancing comparability between the diverse European data sources: the technical development of such indicators is a theme which is taken up again in Part Two.

The more general problems of linking different national census-data series to a variety of geographical data structures are picked up by David Martin in Chapter 4. He describes and evaluates the different geographies which characterize some important national censuses, and identifies some of the uses of census information in a business GIS context. This chapter complements the contributions to Part Two, which all discuss the ways in which census data have been used to create geodemographic information and, equally important, how census data may be used alongside business and service-industry data-sets. David Martin's discussion of population surface models provides one avenue through which the problems of inappropriate and/ or incompatible areal entities raised in Chapters 2 and 3 may be visualized and at least partly addressed.

One of the emerging themes through Part One is that businesses and services cannot take on board the geographical dimension to operational, tactical and strategic planning without first grasping the nature of geographical entities and then understanding how they may be coherently analysed. In short, there is a demonstrable need to understand geographical data structures and the areal units which bound geographical measurements prior to the meaningful analysis of geographical data.

2

Data sources and their geographical integration

David Martin and Paul Longley

2.1 Introduction

The range of data sources that are available to business is wide, and is increasing. This change in part reflects improvements in storage and retrieval of conventional 'public domain' data sources, which increasingly are accessed using powerful yet decentralized computer systems, based for example upon work-station and CD-ROM technology. A range of computer software houses have developed products which enable users to gain rapid access to precisely specified sections of databases, while developments in computer graphics allow information to be portrayed in a geographically accurate manner. Business graphics are, in short, becoming increasingly geographical.

Paralleling these developments is the development of databases by private-sector businesses, ranging from store-card transaction details to electronic point-of-sale (EPOS) and electronic fund transfer at point-of-sale (EFTPOS) data concerning purchasing patterns. Such databases are increasingly sophisticated and, as a partial consequence, are more complex in structure. They are no longer exclusively cross-sectional and/or available only period-ically, as the methods of data capture are equally applicable to monitoring trends and changes in ongoing purchasing patterns as they are for taking 'one-off' cross-sectional profiles of information. Such databases present enormous challenges to database management and analysis, and a recurrent theme of this volume is that this challenge has an important spatial dimension.

Taken together, economic, demographic and social information is becoming available in a much wider and more flexible range of forms than has hitherto been the case. This wealth of data is on a scale unimaginable even only ten years ago (e.g. Wrigley et al. 1985), and yet the sheer magnitude and diversity of data holdings can be quite overwhelming. How can we see the wood for the trees, and create useful information from mountains of data which are rarely directly comparable because of the different ways in which they are structured?

In this chapter, we will suggest that geography provides the 'natural' medium for integration of diverse data sources available to businesses and service providers. Indeed, this recent explosion in the scale and scope of computer-readable databases has been a powerful force in the development and uptake of geographical information systems themselves. At the present time, many business and service organizations are finding that a combination of public and private databases are required in order to fulfil market analysis functions. Such sources are likely, however, to involve different areal units of data collection and availability, and also possibly different geographical coverages. An understanding of the basic ways in which areal coverages may be created, combined and hence analysed is thus a prerequisite to progress.

In order to illustrate the problems of integration, we will develop two extended examples in this chapter. The first examines the ways in which areal units may be managed in order to resolve basic geographical queries. This example presumes use of a single data source and emphasizes analytical operations rather than practicalities. The second example considers some practical problems of geographical data integration when diverse public and private sources are combined, and is very much a 'real world' example. Both examples are concerned with the fundamental problem of geographical analysis, which is that continuous space (the 'natural' medium for data integration) is in practice divided into discrete 'building blocks', which are neither 'natural' for most analytical purposes, nor consistent between often diverse raw data sources. Only when this fundamental problem has been recognized and, for practical purposes, resolved can geographical data be manipulated in order to 'add value' and create useful information (Birkin and Clarke 1988; Thrall and Thrall 1993) for market analysis and other functions.

Many of the themes which will be raised here are explored in greater detail in later chapters: Martin (Chapter 4) explores the ways in which census geographies may be depicted and manipulated in more useful ways; Batey and Brown (Chapter 5) discuss how census-type information has been used to develop fully fledged geodemographic indicators; and Birkin (Chapter 6) describes how diverse data sources may be 'fused' in order to freshen up standard sources with other customer information. The recurrent theme in all of this is that if analysis is inherently geographical, then a clear conception of the nature and functioning of geographical entities is a prerequisite to progress. GIS provides the most appropriate medium for storage, retrieval and analysis of information in these applications, but GIS alone can provide no 'technological fix' for a vague or even misleading conception of the areal entities that provide the fundamental building blocks for geographical analysis.

2.2 Diverse data-sets, different data structures

Geographical location is an important attribute of almost all phenomena, but can be difficult to come to terms with in meaningful analysis. The reason for this is that we are often required to match the characteristics or locational attributes of individuals with various geographical aggregations, such as streets, small census areas, market areas and so forth. In short, the basic building blocks of most market area analyses are aggregations which may not be appropriate to our own analysis, and which may or may not be geographically coincident with one another. Put simply, in geographical analysis we cannot ascribe even the very dominant characteristics of areal data to individuals and point locations within those areas. This problem, known as the ecological fallacy, has long bedevilled geographical analysis, and is one which is never likely to be completely solved in a theoretical sense.

A related problem is that areal units may assume different characteristics at different scales of aggregation. For example, the immediate catchment area of a doctor's surgery may be very deprived, although this concentration of deprived individuals might be diluted when data are presented for larger areas, which include relatively affluent neighbourhoods (Martin *et al.* 1994b). Even the immediate catchment area may not appear to be deprived if the boundaries of the areal units used for data aggregation cause the local concentration to be split between two or more areas. This is illustrated in Figure 2.1, in which the local concentration of 50 per cent (250 of 500 households) deprivation around the surgery site (a) is dispersed across two zones each recording only 33 per cent (250 of 750 households) deprivation scores. More generally, the problems of scale and aggregation are together known as the *modifiable areal unit problem* (Openshaw 1984a; 1984b). Some of the implications of these sorts of problems for spatial analysis are discussed by Openshaw in Chapter 7 of this volume.

Given our limited understanding of scale and aggregation, we nevertheless need to be able to carry out robust and defensible analyses using the available geographical information. Spatial objects are commonly categorized as of point, line, area and surface types (Unwin 1981). These geographical primitives may be captured, stored, retrieved and analysed within GIS using a number of different structures for data organization, by far the most common of which are based on either vector or raster concepts. Some more specialized applications will utilize object-oriented or triangulated irregular network

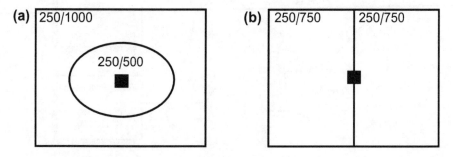

Fig. 2.1   Dilution of target concentration by the geography of data collection units

(TIN) data structures (Burrough 1986, chapter 3; Weibul and Heller 1991), but few business applications currently utilize these approaches. Raster data comprise a mosaic of regularly shaped areal units (pixels), superimposed upon the geographical extent of the study area, which may be used to define features such as points, lines or areal units. Each variable in a raster database is represented by a complete grid coverage. Raster-based GIS function by modification of the cell values in one or more input coverages in order to create new output coverages. Such systems typically include functions for the reclassification of cell values *in situ*, the overlay of values in a number of input coverages, and distance and neighbourhood-based calculations (Berry 1987; Tomlin 1991). Large-scale environmental databases, captured using remote sensing (satellite) technology are commonly of this type. Other sources of raster data include the results of interpolation and surface-building models, and vector data which have been rasterized.

Vector models, by contrast, are based around individual geographical objects which are described by series of coordinates, or grid references. The points, lines and areas associated with socio-economic data are more commonly captured in vector form by digitizing, although much data for business may be modelled adequately in either vector or raster form. The final choice between vector or raster often owes as much to cartographic tradition and software considerations as to any specific properties of the input data. Martin (1991) provides a full description of how topological relationships can be encoded while digitizing strings of points, and hence how digitized data may be structured in order to represent linear and areal objects such as roads and market areas. In vector data models, the geographical and non-geographical attributes of each object are held in separate parts of the database. An example is provided by Figure 2.2, in which coordinate information describing the boundaries of a census zone is linked to the population characteristics found

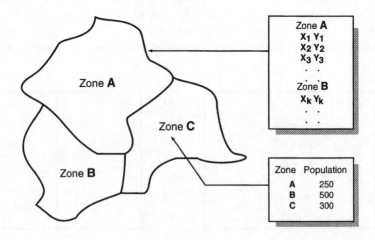

Fig. 2.2 Matching of geographical and attribute information within a GIS

within that zone. In reality, additional attribute information may also be associated with the individual boundary segments or individual digitized points.

The range and scope of analyses that may be carried out using GIS will of course depend upon the sorts of geographical objects about which any given data-set is based. In the United States, for example, data-sets such as the TIGER (Topologically Integrated Geographical Coding and Referencing) files produced to accompany the 1990 census, are based upon lines which form both a description of the street network, and also a set of boundaries for the representation of census block boundaries (see Figure 2.3). These have been produced by the US Geological Survey, working in association with the Bureau of the Census. A street segment or block will typically comprise 250–500 housing units in urban areas. In the United Kingdom, census analysis is structured about small area enumeration district data. A UK urban enumeration district will typically comprise around 200 households or 400 individuals. Figure 2.4 illustrates a typical tiling of enumeration districts which is available in digital form.

Ultimately, as implied above, the only completely disaggregate data would be individual point-level information. Such information is now becoming available in the United Kingdom as the Ordnance Survey Address Point product. Here, every postal address in Britain is being assigned a unique reference code, together with an Ordnance Survey grid reference being digitized to 1 metre precision, and thus address-based information may for the first time be unambiguously georeferenced. Plate 2.1 illustrates the Address Point product. Although more detailed and geographically accurate than other data products available for the United States (e.g. TIGER) and elsewhere, such information cannot, of course, offer ultimate solutions to geographical analyses. The reason for this is that many data sources (and almost all secondary data sources such as censuses) are only made available for aggregations, for reasons of confidentiality, and thus inference from aggregations to individual points remains vulnerable to the modifiable areal unit problem. There is thus never likely to be a purely technological solution to the problems posed by areal aggregations.

The case for structuring market research and service delivery systems using geographical point, line and/or area referencing is a strong one, since a potentially wide range of information sources may be integrated and analysed across the geographical dimension (Department of the Environment 1987). In practice, however, there are often severe problems in analysing different data, because the geographical units and data structures for which data are collected may be quite different. Dwellings, properties and households are the basic units of consumption of goods and services, yet the consuming populations are inevitably referenced indirectly: for example, via census zones, postal geography, property codes or other administrative divisions. One of the main challenges in using GIS in business and service planning is therefore the task of defining a feasible and appropriate spatial referencing framework for diverse socio-economic data sources (Martin *et al.* 1994a).

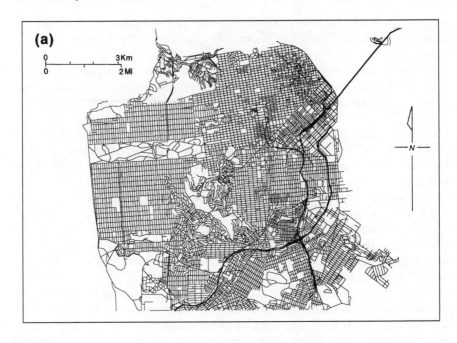

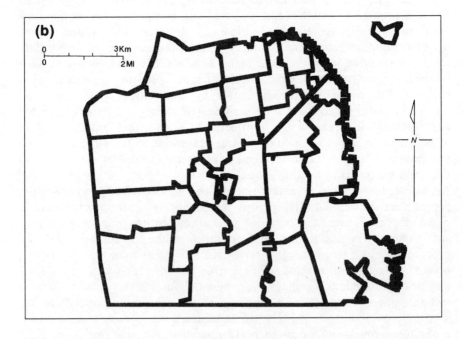

Fig. 2.3   The geography of the 1990 US Census: Downtown San Francisco, showing digitized streets from TIGER files and zipcode boundaries

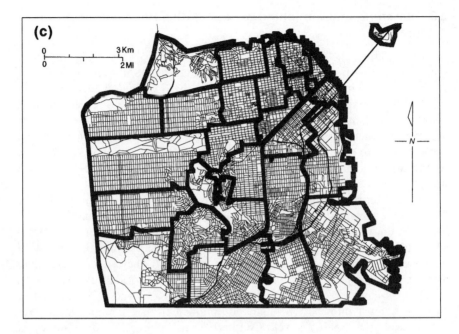

Fig. 2.3   *continued*

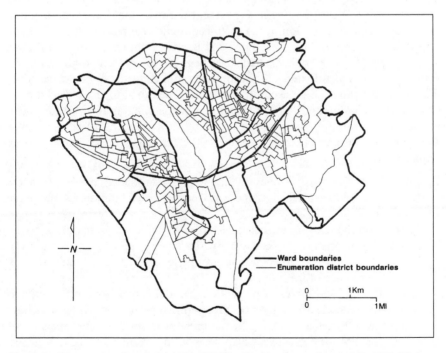

Fig. 2.4   The geography of the 1991 UK Census: the Inner Area of Cardiff, Wales, showing in-house digitized ward and enumeration district boundaries

2.3 Matching and modelling diverse data sources: concept and practice

Martin (Chapter 4, this volume) discusses some of the wide range of geographical questions that may be addressed using GIS, that is the problems of the relationships between areal units and the distribution of data within them. In this section we begin with a hypothetical example of a geographical targeting problem, and then describe our own experience with a messier, yet 'real world' practical problem. The first example will serve to highlight some of the potential for solving geographical allocation problems with GIS, while the second will illustrate the constraints under which 'live' applications must operate.

A typical business or service planning application of GIS might involve the following: a pizza restaurant chain have divided up a city into market areas, based on the travel distances over which it is practical to deliver their pizzas while still hot. (This task in itself would be an interesting application for a GIS with a road database and some networking functions!) They are now seeking to assess the delivery area of a new restaurant in order to identify those parts which most closely match the socio-economic characteristics of their existing customers. Such information might be obtained from a commercial market data consultancy (see Cresswell, Chapter 9, this volume), or by a survey of customers at their other pizza outlets. It is intended that the most promising parts of the new market area be mailed with a leaflet advertising the new delivery service. We might assume for present purposes that information about the wider spatial distribution of the socio-economic group is available at small area level from census data. (See Batey and Brown, Chapter 5, this volume, for a discussion of more sophisticated classifications of customer groups.) We assume also that all addresses in the study area have postcodes/zipcodes, that the unit codes can each be accurately georeferenced (e.g. to the UK National Grid or latitude/longitude in the US) but that the geographical arrangement of these codes is different from the geographical patterning of census areas. The geographical analysis might comprise the following stages, illustrated in Figure 2.5.

1. Identify all the census zones falling within the shaded market area by overlaying digitized census zone and market area boundaries (polygon overlay).
2. Select from the relevant census zones those which have the strongest pizza-eating characteristics, as revealed by the survey data (perhaps younger households, higher socio-economic groups, etc.). In this example zone B meets the required threshold.
3. Use either look-up tables of postcode (zipcode)/census zone intersections, or obtain postcode grid references and perform point-in-polygon search to identify those postcodes falling in the most promising census zones.
4. Using the list of postcodes which lie within the relevant census areas, extract the list of addresses which are likely to house persons of the desired socio-economic group, and mail publicity materials.

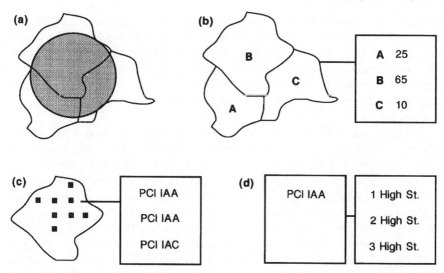

Fig. 2.5 Stages in the matching of geographical boundaries and market attribute information

Of course, it is simplistic to assume that census information would be likely to contain a single perfect indicator of pizza sales potential, for consumption habits do not correspond so directly with membership of a particular socio-economic group. In practice, information from censuses and other sources is combined in order to generate summary indices of behaviour and/or needs (e.g. for a discussion of deprivation indices see Hirschfield 1995), or indeed of propensity to consume different products and services. It is this lattermost function that the geodemographics industry, described by Batey and Brown (Chapter 5, this volume), has developed to serve. Despite simplification, the example cited here is in fact similar in many respects to the methods used by the hospitality industry to determine the most appropriate format for roadside cafes, motels and restaurants, tailoring outlets to best suit the profile of the local population (see Cresswell, Chapter 9, this volume).

Such geodemographic and needs indicators are almost inevitably imprecise, in that no combination of any number of available variables is likely ever to amount to a 'total' explanation of the geographical incidence of a need or a market opportunity. This is because any set of indicators (direct or indirect: see Bracken 1981, pp. 281–5) is unlikely to amount to an exhaustive definition of socio-economic conditions or phenomena. This is widely recognized and accepted in social and economic investigations. Moreover, the addition of the geographical dimension creates additional uncertainty, since the robustness of the spatial matching procedure is crucially dependent upon the success with which point-based addresses can be located with respect to the (usually different) geographies which characterize census and postcode/zipcode zonings. The possibility of mismatch generates a 'fuzziness' to potential findings beyond those encountered in aspatial analyses. As Clarke and Clarke (Chapter 10, this volume) point out, there is nevertheless a strong motivation to reach a practical solution to these problems, since geography is not just an ancillary variable in the pattern of need and demand: rather, it is a fundamental organizing principle, according to which needs and demands are structured.

The practical problems associated with spatial data management will now be illustrated with respect to an actual empirical example. As part of a wider study of housing needs, work for Cardiff City Council (a UK local authority) has sought to target areas of the city for grant-aided housing improvement activity. Cardiff developed rapidly in the late nineteenth century in response to the development of the South Wales Coalfield, and the majority of properties in its so-called 'Inner Area' (shown in Figure 2.4) are over 100 years old. The Area comprises some 47 000 properties, which house the 89 000 adult residents of 920 streets. The predominant built form is terraced housing and, largely because of its age, much of the housing stock would continually slide into disrepair were it not for funded improvement and repair initiatives. In order to identify various attributes of the dwelling stock and the characteristics of its residents, the city council periodically carries out surveys of the dwelling stock, the most recent of which was a major household survey of a one-in-five stratified sample of the private-sector stock (Keltecs 1989). This was an expensive undertaking, but information on some property attributes (e.g. property age) was of long-term utility in assessing needs. Other characteristics (e.g. the distribution of various income groups) were of more transitory importance, but nevertheless could be updated with reference to small area indicator data in later censuses of population. Information on attitude towards funding repairs from household income and/or savings was also elicited.

The geographical problem which arises for the local authority is part allocation, part targeting. On the one hand, an objective is to allocate available monies for grant-related expenditure such that the spend exerts maximum leverage on the overall condition of the built environment (properties and streets) within the city's jurisdiction. This involves targeting contiguous groups of properties which are in a poor state of repair, in order to reap the economies of scale of 'block repair' schemes: this may also unlock additional sources of funding for associated improvements to the street environment. On the other hand, the local authority is required to fulfil a commercial bureau-type function, in that grants should be targeted towards groups of properties where most or all residents are likely to participate in block repair schemes: for example those who are eligible for 100 per cent grants on means-tested grounds and those who are most likely to contribute towards the upkeep of their properties. This combined allocation/targeting exercise is subject to the additional constraints of property age, size (as measured by 'rateable value') of property and duration of residence.

For present purposes we can consider the grant-targeting/allocation exercise to take place under the following circumstances:

1. Properties must be of a certain minimum age, with pre-World War I properties assigned the highest priority if grant demand is high.
2. Grants are intended for single-family properties rather than the very largest properties, which are in multiple occupation, i.e. those with values beneath a ceiling ('rateable') value.
3. Grants are means-tested, so as to target finance towards those households who are unable, rather than unwilling, to undertake repairs and improvements from their own resources.
4. Additional government monies have, in the past, been made available for environmental improvements in residential streets and other small areas where considerable property improvement activity has taken place.
5. Grants must be repaid if the recipient moves from the property within three years of completion of the works.
6. Improvement of a number of adjacent properties at the same time leads to economies of scale relative to the effects of 'pepper-potting' grant aid over a wide area.

The corresponding objectives to these six points in the geographical targeting of grants are therefore to assign lower priorities or to remove from consideration properties which are: (1) of more recent construction, as measured by specially commissioned street surveys; (2) unusually large, as measured by official lists of (non-georeferenced) property (rateable) values; (3) occupied by high income (i.e. non-grant-eligible) households, as ascertained by a specially commissioned survey; (4) not clustered in compact geographical streets or other localities with other eligible properties, since this would remove economies of scale (see also (6) above) and would not create eligibility for further grant-aided street improvements, as identifiable on a digitized street plan; and (5) subject to very high residential mobility rates, as identified by the population census and a specially commissioned survey. The problem is thus geographical, in that various property attributes and household characteristics require sifting within a geographical framework, and the attribute of geographical contiguity at the street scale is itself a central aspect to the targeting problem.

The objective of the exercise is thus to 'filter' successive data-sets in order to identify a subset of small areas in which it would be desirable on grounds of property age and condition to undertake grant-related activity, and then to identify a further geographical subset within which 100 per cent grants would be available or matching funding is likely to be forthcoming. This filtering process involves using a number of diverse and sometimes not explicitly geographical data-sets. In practice there are three main information sources: (1) the domestic rates register for the entire city, to provide property (rateable) value information; (2) the dwelling attribute and household characteristics data-set, which was collected in the major 1989 Cardiff House Condition Survey; and (3) the 1991 UK Census, to provide updates on some household characteristics such as residential mobility rates.

These data-sets are typical of the important sources of information that will be found in the context of any large business or service organization, and we have argued previously (Martin *et al.* 1994a) that GIS presents the most appropriate framework for their integration. Postcodes/zipcodes (and related products) are increasingly being put forward as the most suitable means of cross-referencing socio-economic data-sets (Raper *et al.* 1992), although many data-sets in general use are not (fully) postcoded – as was the case in our own work with the rates register. In particular, the customer referencing systems used by most utilities comprise hierarchical 'Unique Property Reference Numbers' (UPRNs), derived from delivery area and street segment information, but without any explicit requirement for postcoding of records.

Streets, and subdivisions of streets, provide an important intermediate stage in this kind of urban spatial referencing system, although it is interesting to contrast the relative unimportance of streets to date in the UK context with their central role in the highly successful US street-based census geography using DIME (Dual Independent Map Encoding) and TIGER (Topologically Integrated Geographical Coding and Referencing: Cooke 1989). Street segments can be combined to form most levels of zonal hierarchies, such as census zones and other small policy areas. However, street segments have figured little in the debates about basic spatial units owing to the lack of a central directory or set of definitions, and the difficulty of tracing textual information on street addresses, which may be recorded in many different formats. Digitized street centre-lines are often available from mapping agencies (e.g. the US TIGER files and the UK Ordnance Survey OSCAR product), but the pricing of such products varies considerably according to the cost-recovery constraints imposed upon such agencies (Rhind 1992) and to some degree according to the level of topographic and topological accuracy of such products. Alternatively, street networks may be digitized in-house (typically at 1:1250 and 1:2500 scale), although royalties may also be payable in respect of national mapping agency copyrights. Digitized street networks provide a clear link between address-type information and explicit grid references, but really require additional information concerning the address ranges to be found on each block.

For the purposes of our Cardiff application, we have developed what we have termed a 'loosely-coupled' GIS using the ARC/INFO GIS alongside a Lotus 1-2-3 spreadsheet (see Maguire, Chapter 8, this volume). Property value information is held at the street level for the 920 streets in the Inner Area. Data are stored in a Lotus 1-2-3 spreadsheet and linked through a seven-figure code derived from the City UPRNs to the digitized street network in ARC/INFO (Figure 2.6). Aggregation of such data to various administrative units is then possible using the GIS in order to investigate a number of potential policy scenarios (Martin *et al.* 1992). The data are exported from Lotus as an ASCII file, and then in the Tables module of ARC/INFO, a template is defined and the ASCII file is imported into this. The seven-figure street code is the 'relational join' between this file and the digitized street network. Using Joinitem in ARC/INFO, a series of attribute data from the

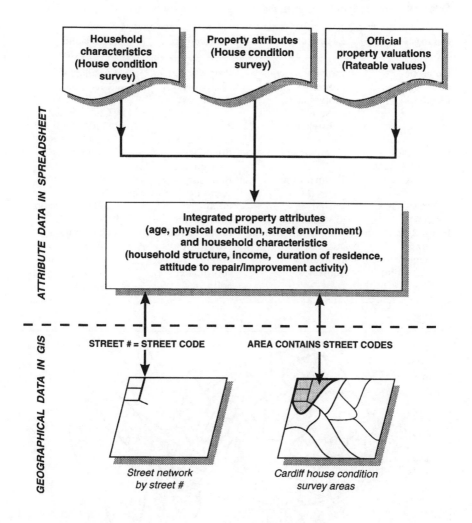

Fig. 2.6 Linkage of spatial and attribute information within a 'loosely coupled' GIS

Council-commissioned House Condition Survey have been attached. The range of information held in the GIS is shown in Table 2.1.

A widespread problem in GIS-based research relates to the portrayal of the results from analysis. Specifically, intra- and inter-organization confidentiality restrictions mean that GIS outputs should maintain anonymity through data aggregation, in order to ensure that individuals with unusual characteristics are not inadvertently made identifiable in the output data. In other data-sets it is necessary to equate street line information with small census tract (UK enumeration district) information. Figure 2.7 shows the tiling of the House Condition Survey areas, devised by the City Council on the basis of known or

Table 2.1 Primary databases available for the Cardiff study

Database	Basic unit	Georeference	Availability
Domestic rates register	Individual properties	Property Ref. No./addresses	Available via City Council
Electoral roll	Individual persons	Addresses	Available via City Council
Streets	Centrelines	Segment coordinates	OSCAR/1:1250 or 1:2500 maps
Uniform areas (e.g. HCS areas)	Small areas	Boundary coordinates	Own survey/ digitizing
Census data	Enumeration districts	Boundary coordinates/centroid	OPCS via MCC
ED/postcode directory	Partial postcode unit	Point coordinates	100 m precision from CPD
House Condition Survey	Sampled individual addresses	Property Ref. No./addresses	Confidential survey
Ordnance Survey Address Point	Individual addresses	Point coordinates/ addresses	1m precision commercially available

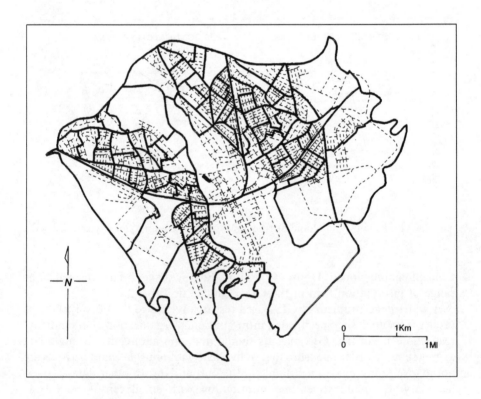

Fig. 2.7 House Condition Survey geography of Cardiff's Inner Area and digitized street network (street network Crown Copyright)

likely within-area homogeneity of built form, which are quite different from the 1991 census enumeration district boundaries (Figure 2.3). Figure 2.7 also shows the digitized street pattern. It was not generally possible to georeference properties within streets, and thus the street is the finest level of resolution in the analysis. Where it was necessary to assign street information into small areas such as the House Condition Survey areas, a line-in-polygon analysis of the entire data-set was performed in order to identify those street segments which crossed polygon boundaries, and the attribute information from these segments was then divided between the polygons concerned. A line-in-polygon search involves the geometric calculation of intersections between each line and polygon boundary, and allows numbers of properties to be assigned to areas in proportion to street length in each area. Similar problems arise when attempts are made to reconcile census information with street aggregations, particularly since tract (enumeration district) boundaries frequently run down streets and across junctions, making the assignment of houses and their occupants into tracts ambiguous.

The linked database does, however, make it possible to begin to filter out likely areas in which to target housing grants on the basis of demonstrable need, likely uptake and economies of scale. The GIS in effect creates a number of new data entities, and analytical outputs have been possible by spatial integration of the data which could not have been performed using the existing databases in isolation or by simple cross-matching of lists. Indeed the difficulty of list-matching for statistical purposes could largely be overcome by spatial manipulation of the data, although the full potential for this will only be achieved with a combination of individual property grid referencing and unique property reference numbers, as considered below. Figure 2.8 shows how this geographical matching process can be used to identify those street segments in which household characteristics and property attributes are the most conducive to grant-aided improvement activity based upon constraints (1) to (6) above.

2.4 Diverse data sources in practice

Our experience is illustrative of many of the difficulties which beset any attempt to use existing large property databases within GIS. Many of the issues which we have encountered here echo those raised by Brusegard and Menger (1989). These may be considered under two headings: those attributable to data sources, and those associated with analytical operations.

Any attempt to handle property registers on this scale is bound to be beset by problems relating to overlaps and mismatches of various kinds. Each time a new register is included in the base for analysis, the number of properties which can uniquely be traced in every register is further reduced. Such problems are particularly acute in areas undergoing change in built form, for example through redevelopment or property subdivision (the latter generating particular problems of clearly matching household and property information,

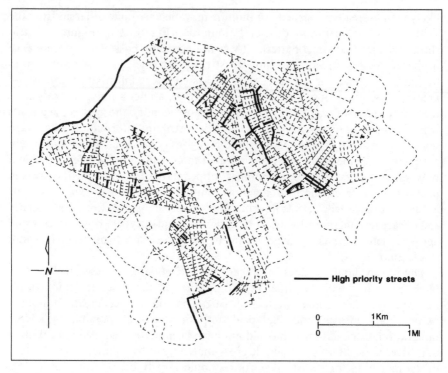

Fig. 2.8 Result of filtering household characteristics and dwelling attributes in order to target housing improvement grants

especially in areas in which individual properties are occupied by more than one household).

The lack of grid referencing at the level of individual properties, and the need to assign parts of street segments into arbitrary areas, inevitably lead to a degree of 'fuzziness' in the locational referencing (Openshaw 1989a). The data problems we face provide good examples of most of the classes of 'obvious' error identified by Burrough (1986), as shown in Table 2.2. The usual survey research problems are no less relevant to GIS-based research, notably non-response among survey samples and omissions of individuals who manage to evade even census-type surveys and registers. GIS are frequently reliant upon secondary data sources, and careful consideration must always be given to the quality and purpose of the original data-collection exercise, in the context of the analysis being performed.

In addition to the (relatively) simple problems inherent in the data-sets which we have been using, there is an additional class of problems which are associated with the analytical operations which we have performed. The GIS has been used in order to provide linkages between data-sets based on spatial coincidence, where no other mechanism for linkage was available. This has created difficulties which are due entirely to the technique being used, and are

Table 2.2 Examples of error sources as originally identified by Burrough (1986)

Age of data:	Address registers which predate recent demolition and construction
Areal coverage:	Difficulty of obtaining price survey information for areas of primarily local authority-owned housing
Density of observations:	Again, varying density of house price information, as determined by the property market and availability of estate agents' information
Relevance:	Need to use electoral register as a surrogate for the community charge register, which is confidential and hence unavailable
Accessibility:	Confidentiality concerns regarding council tax register, and housing condition survey information
	Cost implications of OS copyright, OSCAR and Pinpoint Address Codes (which we could not obtain)
Positional accuracy:	Digitizing error in HCS areas and street centrelines
Accuracy of content:	Obvious errors in textual address fields
	Hard to assess (e.g.) accuracy of rateable values

unrelated to the substantive research problem. The line-in-polygon analysis for allocation of addresses in street segments provides a unique GIS-oriented approach to the matching of database entities, but is made unreliable because of the resolution of the input data.

An entirely different set of issues relates to the difficulties of data representation, when dealing with address-referenced information. When the basic entities are points (addresses) and lines (street segments), presentation of complex patterns in the data is difficult, and we have resorted to the use of small multiples (Tufte 1983). This problem will become increasingly manifest as the more routine use of point-based data-sets (such as the UK Ordnance Survey's Address Point product) permits precise identification and targeting of points, while confidentiality restrictions prevent the display of such disaggregate information in reports and secondary analyses. Plate 2.1 illustrates how Address Point can be used to highlight the properties that were sampled in a housing survey, but would violate confidentiality constraints were any household characteristics assigned to the sampled points. In our own work on this data-set we have needed to assign modal values to street segments for display purposes. The GIS environment offers no adequate tools for the depiction of relationships between variables at the individual property level, and aggregation is almost always necessary simply in order to record results on screen or paper. In some cases, aggregation is also imposed by the requirement for confidentiality, and census-like restrictions on small numbers are necessary. These restrictions are frustrating in a context where there is the opportunity to work with detailed data-sets, and to explore complex relationships at the level of the individual household.

Prospective GIS users should be aware that at various stages of conducting data-rich geographical research it may not be possible to perform statistical analysis or list-matching required directly within the GIS, but rather it will prove necessary to export the information to more general statistical packages or even to write customized programs – this is increasingly possible as the scope of analysis becomes more explicitly geographical and information systems become more customized than generic (Cresswell, Chapter 9, this volume; Clarke and Clarke, Chapter 10, this volume; Maguire, Chapter 8, this volume).

2.5 Concluding comments

There is increasing evidence that GIS has 'come of age' and that it now has a significant role to play in both specialist and routine business applications. Many of the outstanding difficulties in applications relate to data problems which limit the analyses which are possible, or to the lack of appropriate tools within the GIS for tackling our *ad hoc* statistical and modelling needs. However, the development of more complex spatial analysis using GIS tools is less of a constraint than those issues relating to the data.

Data issues are therefore of utmost importance, because of the complex processes of error propagation which occur during analysis of such data-sets. Data on properties, households and individuals are a fundamental form of geographic information, and their use is growing. One of the biggest problems with such data sources is (and will continue to be) the proportion of mismatches which occur as successive thematic layers are added to the analysis. Our own experience suggests that the solution lies either in national detailed georeferencing, or in a series of widely recognized hybrid units such as street segments, which would provide a form of 'fuzzy geography' for the handling of these kinds of information. In order to be of assistance, national georeferencing systems must be sufficiently widely available (in terms of cost, data formats, etc.) to make their integration with existing lists a viable option. Cost recovery and the commodification of information are crucial issues here (Waters, Chapter 3, this volume; Openshaw and Goddard 1987), as are unresolved issues of information copyright and data standards.

3

Data sources and their availability for business users across Europe

Robin Waters

3.1 Introduction

The needs of businesses for data and information are universal. In Chapter 2
Martin and Longley discussed some of the 'technical' difficulties associated
with assembling information systems based on diverse data sources. The aim of
this chapter is to review the general availability, cost and ease of transfer of the
types of data which businesses may need to integrate for successful pan-
European operations. It should be recognized at the outset that the availability
of such data and information can be markedly different from continent to
continent and from country to country. Obtaining, integrating and interpreting
these data in any one country is not easy; for pan-European operations it may
be very difficult indeed, for the following reasons:

1. Data may be missing – not all countries collect the same data.
2. Data may be confidential – data protection is a very sensitive political
   issue.
3. Data may be collected in different ways – e.g through censuses (which
   may themselves concern an entire population or just a sample of it) or
   through address registrations.
4. Data may be collected for different areal units even within a single
   political jurisdiction: for example, within the UK, general-purpose
   postcode units are used as census building blocks in Scotland while
   census-specific enumeration districts (EDs) are used in England and
   Wales.
5. Data are very variable in cost – the costs of undertaking censuses and
   other large-scale surveys are high, and different countries have different
   cost recovery policies. This in turn affects the price of 'value added' data
   products.

6. Data vary in their currency. Any conventional census-style data set is but a 'snapshot' of conditions at a point in time, and there is no coordination of the frequency or date of data collection across different countries.
7. Data vary in reliability – data collected by post may be less reliable than that collected in face-to-face interviews.
8. Censuses and surveys vary in content – in part because different data collection methods are amenable to a more or less restricted range of questions (for example, sensitive issues of income or ethnicity are rarely broached successfully in postal surveys).
9. Cultural factors influence the availability of data and their nature.
10. Military attitudes can influence the availability of topographic data.
11. Religious attitudes may influence the collection of personal information.

We shall explore a number of these issues in more detail below. In the next section we review the agencies involved with providing basic cartographic information which might form the base for any GIS. In Section 3.3 we describe the availability and costs of population, economic and geodemographic information both at the pan-European level and within individual EU states. The availability of both cadastral and statistical information and the ease of data transfer is discussed in Section 3.4. Some brief concluding remarks are offered in Section 3.5.

3.2 European data sources and the role of national mapping agencies

The national mapping agencies (NMAs: see Appendix A) do not, in general, provide multinational products, although there is usually some overlap at national borders. Rather, the national mapping agencies produce series of maps and data-sets which vary considerably from country to country in terms of scale, content and currency. All have traditionally been funded from the public purse to a greater or lesser extent. CERCO (Comité Européen des Responsables de la Cartographie Officielle) is the NMA's joint organization which tries to coordinate content, standards and currency of topographic detail and administrative boundary information. Because of the influence of the military, on both sides of the former Iron Curtain, there is a relatively consistent set of maps available at 1:50 000, 1:250 000 and smaller, and these show topography, as well as administrative boundaries and the names of both natural and man-made features. The extent to which these maps have been turned into computer-readable data-sets is, however, much less consistent. The contours, or at least terrain models derived from them, are available across the whole of Europe to a consistent standard which was designed for various military purposes. However, with the exception of the siting of radio transmitters and receivers, there are relatively few commercial users of pan-European terrain data at this level of detail.

There are signs that this is changing, however. CERCO was set up by the heads of Europe's national mapping agencies in order to create a forum for activities that required cross-border cooperation. CERCO is not an agency of the European Union and includes over twenty countries. It has recently established the 'MEGRIN Group' (Multi-purpose European Ground Related Information Network) with the aims of promoting the NMAs to governments, of setting standards for meta-data (information about their products) as well as data standards themselves, and of producing a harmonized set of administrative boundaries in digital form for use in GIS.

In some countries the 1:50 000 maps mentioned above are available as raster-scanned images so that they can be viewed on a screen as a background for other data-sets. However, these can be expensive: the cost of the complete data-set for, say, Great Britain (£40 000 plus annual usage charges) is some 40 times the cost of the equivalent paper maps and is prohibitive for many potential users. At a single-sheet level it is even more expensive, and enforcement of the UK Government copyright means that the commercial sector (or even individual users) may be unable to exploit new GIS technology with the most appropriate data-sets and may thus be forced to turn to lower-resolution alternatives. This is despite the fact that desk-top technology enables less scrupulous users to scan all of the paper maps (cost approx £1000) for a few hundred pounds. Such a discrepancy between official prices and the cost of a workable, if illegal, alternative is a recipe for considerable friction in the future.

At small scales the road centre-lines for all main routes are now available across the whole of Europe from several different suppliers and have been built into PC-based route-finding packages, drive-time analysis programmes and site-optimization systems. For larger scales (as required for in-car navigation systems) there is once again a marked variation in availability of data by country.

François Salgé, the Executive Director of the MEGRIN Group, has roughly categorized the European Union's national mapping agencies by breadth of responsibilities and by degree of commercialization. The resulting matrix (Table 3.1) shows how difficult it could be to reach any consensus on exploitation of maps or map data across Europe.

This categorization is, of necessity, simplified. Germany in fact does not have a 'National Mapping Agency' at all and most of its maps are made by publicly funded organizations in the individual Länder. The UK and Ireland have a tradition, dating from the middle of the nineteenth century, of nationally organized large-scale mapping, which is used as the basis of Land Registration. Although run by the army right up until the 1970s, OSGB now has to recover more of its costs than any other mapping agency in Europe. However, it is also the agency which has provided large-scale digital data coverage first. Researchers may wish to speculate whether or not these features are correlated! Table 3.2 provides a summary of the relative status of national mapping efforts in the different European countries.

In many countries the larger-scale, more detailed maps and plans are the responsibility of local administrations, and this means that they are often of

Table 3.1 Relative status of national mapping agencies in the EU

	Public organization with important commercial activities	Public organization entirely in the civil administration	Military organization or public organization with strong military links
Topographic information at all levels of detail/scale	UK (OSGB & OSNI) Sweden* Norway* Finland* Denmark*	Ireland Germany Austria Portugal Luxembourg	
Medium- and small-scale mapping only	France Spain	Switzerland Belgium	Netherlands Italy Greece

* Note that in Scandinavia, and to a certain extent in the UK as well, the NMAs also have some responsibility for the cadastre or its equivalent.

Table 3.2 Relative status of national mapping in European countries

Country	Area (1000 km^2)	Pop. (M)	Pop. dens (p.p. km^2)$^+$	National mapping and digital status
Austria	83.9	7.660	91	1:50 000 line maps, 230 sheets, digitizing planned 1:25 000 line maps, not digitized 1:10 000 ortho-photo maps, not digitized 1:1000 cadastral maps, digitizing in progress
Belgium	30.5	9.948	326	1:50 000 line maps, scanning and vectorizing in progress 1:25 000 line maps, obsolete 1:10 000 line maps, digitizing in progress 1:10 000 ortho-photomaps
Denmark	43.0	5.135	119	1:25 000 line maps, 405 sheets, all scanned on CD-ROM and conversion to vector data under way. Large-scale cadastral maps being digitized
Finland	338.1	4.998	15	1:50 000 line maps, southern half of Finland 1:20 000 line maps, being scanned for CD-ROM Digital lakes and contours available Large-scale cadastral maps half digitized
France	549.0	56.304	103	1:100 000 line maps, BD Carto data in progress 1:25 000 line maps, BD Topo data in progress Larger-scale maps from local authorities

Table 3.2 contd.

Country	Area (1000 km^2)	Pop. (M)	Pop. dens (p.p. km^2)[+]	National mapping and digital status
Germany	357.5	79.346	222	1:50 000 line maps 1:25 000 line maps, digital data (DLM) in progress 1:5000 line maps, digital data (DGK 5) in progress (All of these are responsibility of Länder)
Greece	132.0	10.046	76	1:100 000 line maps, digitizing in progress
Ireland	70.2	3.506	50	1:50 000 line maps, digitizing in progress
Italy	301.3	57.576	191	1:50 000 line maps, digitizing in progress
Luxembourg	2.6	0.378	145	1:50 000 line maps
Netherlands	41.5	14.893	359	1:50 000 line maps, scanned in colour
Norway	323.9	4.233	13	1:50 000 line maps, digitizing in progress 1:10 000 line maps, digitizing in progress
Portugal	92.1	10.337	112	1:50 000 line maps, digital experiments 1:25 000 line maps, 640 sheets, digitizing in progress 1:10 000 ortho-photo maps 1:2000/5000 cadastral maps, some digitizing
Spain	504.8	38.924	77	1:50 000 line maps, basic digitizing completed 1:50 000 SPOT ortho-images 1:25 000 line maps, not digitized Larger-scale maps are provincial and city produced
Sweden	450.0	8.527	19	1:50 000 line maps Large-scale National Land Data Bank is digital and on-line
Switzerland	41.3	6.674	162	1:25 000 line maps, digitizing in progress Large-scale cadastral maps by city, some digital
UK	244.1	57.409	235	Great Britain (England, Wales and Scotland) 1:50 000 line maps, 204 sheets, scanned in colour 1:25 000 line maps 1:10 000 line maps, scanned in black and white 1:1250/2500 vector database, completely digital in 1995 Northern Ireland 1:50 000 line maps, scanned in colour Large-scale database in preparation

[+] p.p. km^2: people per square kilometre
Source: Eurostat and CERCO (1990).

different standards and currency even within a single country. Moreover, they are often on local coordinate systems, which may not have straightforward relationships to the national or international systems used for medium- and small-scale mapping. These differences have not been important until the advent of worldwide satellite navigation systems. The GPS (Global Positioning System) can now produce positions relative to a world coordinate system to an accuracy of a few metres in virtually real time. However, conversions of these coordinates to national or local grid systems are only now being satisfactorily achieved, and computations often reveal large discrepancies attributable to the less accurate measurement systems and less comprehensive computing facilities used in the past.

Where maps have not yet been converted to digital form, the authorities responsible may now be considering the conversion of the maps to a new, international coordinate system. And yet the routine digitizing of a map does not of course necessarily result in a quality data-set for any particular purpose. For many years map producers digitized their maps only for internal purposes – to produce more maps more quickly. These practical considerations may have generated cartographic products which are not suited to the needs of GIS dedicated to the querying and analysis of single or combined data-sets. In the 1990s we are finally seeing the economies of multiple use of data-sets where data capture costs have been written off and the emphasis is now moving to reprocessing and updating.

3.3 Population, economic and geodemographic data in Europe

3.3.1 Pan-European data

Before addressing the availability of census and sample population sets within different EU nations it is useful to provide an overview of what is available at the pan-European level. GIS users requiring consistent geodemographic data across Europe are in a much more difficult position than their counterparts in the USA. Thrall and Smersh (1994) provide a very useful review of US-wide data-sets that are available from the Federal Government. This information is viewed as a 'National' asset, which is made available on a marginal cost basis, and is the envy of users operating in any individual European country, let alone those operating on a pan-European basis. Appendix B contains a summary of US data agencies.

The European Union's Statistical Office, Eurostat, does its best to ensure some level of consistency within the member states of the EU, and produces some harmonized statistics. However, it has no remit to collect data itself and relies on the statistical institutes of the member states for raw data. Eurostat has set up a spatial zoning system across Europe at five distinct spatial scales. The entire system is labelled the 'Nomenclature of Territorial Units for Statistics' (NUTS). The five distinct spatial scales vary from country level

(NUTS1) to individual towns or communities (NUTS5). The system is based on a nested hierarchy: NUTS1 units consist of a whole number of NUTS2 regional units which, in turn, are made up of a whole level of NUTS3 regions. So far, Eurostat has only been able to publish reasonably consistent information down to the NUTS3 level.

Efforts to encourage individual countries to use similar data-collection methodologies or even to synchronize their data-collection timetables have not met with much success. However, Eurostat is now attempting to build a NUTS5-level database of demographic and economic statistics called SIRE, and it is cooperating with the EU-supported IMPACT project called Euripides, which has similar aims. The fruits of this collaboration will be published commercially.

One of Eurostat's aims is stated to be 'disseminating general European statistical information to assist decision making within businesses and other organisations concerned with economic and social matters'. The geographically related data-sets held by Eurostat are:

1. NUTS boundaries (available in paper or digital forms).
2. Economic and social data. There are some 80 different statistics in the associated REGIO data-sets, which themselves make up most of Eurostat's information base. These are grouped into categories of demography, economics, transport, agriculture, industry, and employment. The spatial resolution of this information base is fairly coarse, although there is often a considerable level of disaggregation. In the demography section, for example, data are available for total populations by age and sex, broken down into 5-year age cohorts. Information on standard birth and death rates is also available. Local economic indicators include total employment broken down by industrial sector, gross domestic product (GDP) indicators, salaries and 'gross value added'.
3. Various background topographical data-sets in association with 'GISCO' (Geographic Information System for the Community). These have been derived from a variety of governmental and commercial data-sets.

Eurostat is often frustrated by the jealously guarded interests of national agencies (for both statistics and mapping), not only for the copyright on their data in itself but also for the different ways in which they are collected. It is unlikely that direct initiatives to encourage consistency in the collection of statistics would resolve this situation. However, changes could be made in other legislation, particularly with regard to EU-funded initiatives, and this could lead to harmonized statistics as a useful by-product. The US precedent of 'matching Federal funds' provides a possible role model, whereby Federal aid for individual state projects is available to match state expenditure, but only on condition that specific data-collection exercises are carried out.

For the business user of GIS there are several other EU-funded programmes which are having an influence on the use of GIS, particularly with regard to pan-European operations. Such programmes include IMPACT2 and COMETT. IMPACT2, run by the DGXIII-E office in Luxembourg, is a 'near market' support programme for new information technology. It has several strands, including one that is devoted to GIS data products. During 1993 it supported a 'definition' phase of 28 projects, covering a range of activities from statistics to height models, from North Sea databases to Mediterranean olive tree information systems, and from educational GIS to tourist multimedia projects. During 1994 it is supporting eight of these projects through their implementation phases to final products. Support in this context means up to 50 per cent funding for all of the research, development and marketing expenses associated with the projects. COMETT is the Community programme for Education and Training in Technology. The programme was launched in 1987 with the primary objective of providing Europe with highly qualified scientific and technical staff. COMETT supports a wide range of projects by: organizing intensive courses on advanced technology; exchanging staff between universities and business enterprises; developing specialized training materials; and analysing training needs at regional level or in a particular technology area.

3.3.2 Population data for individual EU countries

Table 3.3 summarizes the availability, type of count (census every 10 years or more frequent registers), areal units and pricing policy of demographic data across the EU. The 'pricing policy' in Table 3.3 deserves some extra comment. Some national statistical institutes have commercial or semi-commercial status and have to recover some of the collection costs of statistics as well as their processing and dissemination costs. 'Marginal pricing' indicates that an institution has a policy of recovering only the cost of providing the data from the customer. This may still give rise to large differences because of the media required, the amount of data requested and the systems in which the data are held. The 'data access policy' also differs where there are marked differences in the attitude taken to personal data by different countries. In some cases anonymized records for individual households can be made available, provided that there is no way of attributing them to the individuals concerned. The extent to which data protection or other laws affect information availability is also variable.

3.3.3 Public- and private-domain data for market analysis

Different indicators obtained from raw census data are often combined to produce classification indicators for single regions or areas. These are known as geodemographic classifications and are provided by a range of private-sector institutions for business and service planners. They have been central to the development of business applications of GIS, and Part Two of this book is devoted to their history, design and critique. Here, the concern is simply to outline the availability of such systems across Europe.

Table 3.3 Availability of official demographic data in Western European countries

Country	Date	Type of count	Area name	No. of areas	Av. pop. ('000)	Av. area (km^2)	Pricing policy
Austria	1991	Census	Gemeinde	2 333	3	36	Marginal
Belgium	1991	Census	Communes	596	17	51	
Denmark	1991	Register	Kommuner	276	19	156	
Finland	1991	Register	Municipalities	445	11	759	Marginal pricing
France	1990	Census	Communes	36 545	2	15	Marginal pricing
Germany	1991	Surveys	Gemeinde	16 147	5	22	By Länder
Greece	1991	Census	Demoi/Koinotites	6 039	2	22	Marginal pricing
Ireland	1991	Census	DED Wards	3 438	1	20	Some cost rec.
Italy	1991	Census	Comuni	8 097	7	37	Marginal pricing
Luxembourg	1991	Census	Commune	118	3	22	Free
Netherlands	1991	Register and survey	Gemeenten	702	21	59	Some cost rec.
Norway	1991	Census and Register	Municipalities	440	10	736	Marginal pricing
Portugal	1991	Census	Concelhos ou municipios	305	34	302	Some cost rec.
Spain	1991	Census	Municipios	8 056	5	63	Marginal pricing
Sweden	1990	Census and Register	Kommuns	284	30	1 585	Some cost rec.
Switzerland	1991	Census	Communes	3 021	2	14	Marginal pricing
UK	1991	Census	Wards	10 970	5	22	Some cost rec.

Note: Countries with register-based counts will typically have annual data available; those with censuses may also make annual estimates.

The marked difference in the availability and use of geodemographic analysis across Europe can be explained by reference to the different small area zoning systems for which there are publicly available data, as well as the details of the statistics that have been collected. Typical zones are census enumeration districts, unit postcodes, administrative districts and street sections. The relationships between these different zoning schema are not necessarily straightforward. Whereas enumeration districts are usually subdivisions of administrative areas, postal coding and street addresses very often have completely different origins. Since the latter are far more useful for marketing and sales applications, it is necessary to develop methods of allocating appropriate statistics from one set of zones to another (see Martin and Longley, Chapter 2, this volume).

Table 3.4 describes some of the private-sector neighbourhood classification systems that are now available in different countries. Only CCN, with various aggregations of its MOSAIC system, has managed to include more than two countries in its offering (see below). The current releases of some of these, as well as other, geodemographic systems are described in Batey and Brown (Chapter 5, this volume), Birkin (Chapter 6, this volume) and Cresswell (Chapter 9, this volume).

Within Europe, the UK and the Netherlands have the best developed geodemographic systems, with Italy, France and Sweden being next. Finland and Norway have the lowest level of development. It is interesting to note that the UK and Netherlands both have very detailed postcoding systems but have very different data collection methods. The UK has a traditional decennial census (most recently carried out in 1991), but the Netherlands relies on a combination of interviews, telemarketing and official registration.

The aim of each of these systems is to classify neighbourhoods into certain categories which enable the users to make valid assumptions about the likely responses of those classifications to certain consumer goods, to advertising and to services available in the vicinity. MOSAIC has, for example, used 39 classes in Germany. CCN Marketing have now produced a EuroMOSAIC classification across most of Europe, based on 10 lifestyles. EuroMOSAIC is currently the best known attempt at a European-wide neighbourhood classification. It provides consistent geodemographic information for the UK, Belgium, the Netherlands, Germany, Norway, Spain and Sweden. Even MOSAIC, however, differs from country to country in terms of method of data collection and type of basic area (postcode, census district, building). Yet it introduces, for the first time, a single currency for geodemography in Europe, with the following classifications:

1. Elite suburbs
2. Average areas
3. Luxury flats
4. Low income inner city
5. High rise social housing
6. Industrial communities

Table 3.4 Private sector neighbourhood classification systems

Country	Vendor	Geodemo. class.	Geog. unit	No. of units	Average households	Data sources
Belgium (& Lux.)	Sopres/ CCN	MOSAIC	Street	167 000	16	Census Address
Finland	Kohtisoura	Acorn	Census district	n/k	n/k	Census
France	Line Data Coref	Ilot Type	Census district	180 000	130	Census
Germany	Infas	Local	Market res. dist.	60 000	370	Mkt res
Germany	Bertlesmann	Regio	Market res. dist.	60 000	370	Mkt res resp.
Germany	CCN	Micro- MOSAIC	Building	11 million	2	Field svy & phone
Italy	Seat	Cluster	Census district	n/k	n/k	Census
N'lands	Geomarket Profiel	Geo	Postcode	380 000	16	Mkt res
N'lands	CCN	MOSAIC	Postcode	380 000	16	Mkt res Address Mail ord.
Norway	NMU	MOSAIC	Census district	30 000	35	Census
Spain	Bertlesmann	Regio	n/k	n/k	n/k	Mail resp.
Spain	CCN	MOSAIC	Census district	41 000	450	Census Address
Sweden	Marknads Analys	MOSAIC	Post- number	9000	540	Census
UK	CACI	Acorn	Census district	151 000	150	Census
UK	CDMS	Super- profiles	Census district	151 000	150	Census
UK	Infolink	Define	Census district	151 000	150	Census Credit
UK	CCN	MOSAIC	Postcode	1.4 million	16	Census Electoral Credit

Note: details of systems for Austria, Denmark, Greece, Ireland and Portugal were not available.

(Adapted from Watts 1994)

7. Dynamic families
8. Low income families
9. Rural/agricultural
10. Vacation/retirement

Birkin (Chapter 6, this volume) describes some of the characteristics of the EuroMOSAIC product, and lists some of its possible shortcomings: Plates 6.1 and 6.2 depict the distribution of EuroMOSAIC categories within Amsterdam and Madrid, respectively. In addition to census or other 'demographic' data, there are many other sets of statistics that may be available for refining basic geodemographics and/or for keeping them more up to date than is possible with a ten-yearly census. This is discussed in greater detail by Birkin (Chapter 6, this volume).

3.4 The commercialization of data and their availability on CD-ROM

Until the early 1990s the major national mapping agencies and the national statistical institutions were the only suppliers of geographical, demographic and economic statistics. Given the volume of the data-sets involved and the size of computer necessary for analysis or processing of these, the vast majority of data transfers were made by reel-to-reel magnetic tape and as 'raw' data with very basic levels of coding or extraction of subsets. The prices charged were somewhat arbitrary with no consistent policy of marginal costing, full cost recovery or market-oriented pricing.

With the increasing use of PC-based systems (see Cresswell, Chapter 9, this volume), the development of specific GIS and other software for the analysis of geographic and demographic data, and the increasing popularity of the multi-media CD-ROM, CD-ROM is now the most common medium for transfer of census data. CDs are also becoming popular for map information. Furthermore, the pricing policies of some of the major agencies have stabilized, and in some cases the private sector are able to license the raw data. This is leading to the development of 'value-added' data products by value-added resellers (VARs), in many different formats, for a variety of uses and with appropriate pricing (subject to royalties for some of the data suppliers). Many of these private-sector VARs are able to add value by integrating different data-sets, by aggregating statistics to different areas of use for a particular market and by providing software for the users to carry out their own analyses without having to buy separate software packages.

Those government departments that already have aggressive cost-recovery targets and which already provide a range of 'in-house' products are naturally concerned about competition from the value-added sector and seek to enforce very restrictive agreements with the VARs, often by imposing very strict controls on the use of the data by the end user. This is in contrast to the situation in the United States, where Government-produced data are made available to the private sector at marginal cost and the data are wholly in the public domain.

So far, the development of the market for mapping and statistical data has not been characterized by very innovative use of even the existing hardware and software facilities. However, this is likely to change in the next few years. At present, pricing of data-sets is relatively rigid and is not designed to expand the overall market or to address different niches. Differential pricing is available for educational or non-profit research, and some discounts may be available for volume. However, relatively little data are 'shrink-wrapped' or sold on media (such as CD-ROM) which enable different data-sets to be unlocked after a credit-card payment for a code word or number. The encryption technology is now available to make this type of payment relatively straightforward and very secure. Complete software suites from major suppliers are now routinely distributed with magazines on CDs and only require access codes to become down-loadable. In addition to the simplicity and well targeted marketing that this technique provides, it also enables the publisher to maintain excellent records (because a phone-back number is required) and to lock the software/data to a single computer if necessary.

In France there are some services available on line, on Minitel. IGN, the French national surveying and mapping agency, will, for example, provide coordinates for its reference points around the country and the charges may be billed to regular telephone accounts. Although limited by the 'tele-text' format available on the Minitel service, the facility is well used for the basic searching and transfer of small amounts of relatively high value information. In the UK, Phonelink raised capital with a stock market floatation in 1993 and their shares rose dramatically even before the launch of their service – Tel-Me – in June 1994. Tel-Me provides an on-line dial-up service for Windows users giving access to phone number directories, railway timetables, commercial directories, up-to-the-minute news and weather, national mapping, hotel directories and company performance. Charging is made through an initial joining fee (which includes the software for the user's PC) and then subsequently according to connect time for each of the different services.

It is clear that most users of mapping and statistics databases require either summaries of large areas or more details of smaller areas. There is a trade off between depth of analysis and geographical area to be covered. There is also a premium available for the provision of the most up-to-date information. This will lead to a division of data provision with relatively static information on CD-ROM and with an additional service for more up-to-date information on-line. Table 3.5 lists some private-sector data suppliers within the EU. It is also necessary to provide information in user-definable chunks without increasing distribution and handling costs. This is possible either with shrink-wrapped disks with unlockable data or with interactive databases accessible over value-added networks with charging facilities.

Given that we are currently in an early stage of market development of data for GIS, it is vital that experimental products and services should be encouraged without too much regulation. The EU has used its various 'high

Table 3.5 Private-sector data suppliers by country

Country	Companies
Austria	WIGeo-GIS GmbH
Belgium	Tele Atlas
Denmark	Unknown
Finland	Unknown
France	GEO.RM, SPOT Image
Germany	Institut für Rauminformationen (IRI)
Greece	ERATOSTHENES, OMAS, EPSILON
Ireland	ERA-MAPTEC
Italy	Touring Club Italiano (TCI)
Luxembourg	Unknown
Netherlands	Bureau Nieuwland, Geodan, European Geographic Technologies (EGT), AND
Norway	Unknown
Portugal	Octopus, Proambio-Projectos de Ambiente
Spain	INFOCARTA, MAPTEL
Sweden	Satellitbild
Switzerland	Intersurvey
UK	Bartholomew/Times, Automobile Association (through Kingswood) MVA Systematica, MR Datagraphics, Chas E Goad, GDC, The Data Consultancy, GeoInformation International

From *GIS Sourcebook* (GIS World 1993) and author's own research.

tech' support programmes to help some pan-European projects. In particular, the IMPACT2 GIS programme (see Section 3.3.1) is supporting eight projects with funding of nearly 300 000 ecu each. These were chosen from a range of thirty projects funded to a definition phase. The final eight include small area statistics, geographic meta-data, offshore databases, education and travel information projects.

3.5 Conclusions

From the preceding discussion, it is clear that there is a richness and diversity of data-sets and products available for countries in the EU, and that these hold enticing prospects for pan-European GIS-based analysis. The limited development of such analysis to date nevertheless illustrates that there are a number of problems which must be overcome before meaningful geographical analysis might be carried out at scales commensurate with internationalized business and service operations.

First, and fundamentally, the quality and quantity of business-relevant data varies considerably within the federal EU trade area, and this creates problems for geographical analyses. Large-scale public-sector surveys in different countries actually measure rather different population characteristics, and it thus becomes difficult to devise consistent and coherent pan-European indices of the propensity to consume different products and services. The discriminatory power of such tools is blunted as a consequence. Moreover, the social and cultural factors alluded to in Section 3.1 may create further international variations in the correlations between consumption patterns and socio-economic and demographic characteristics. The sort of raw data that are collected very much determine the range and scope of analysis that can be carried out for any area, and there is thus a vertical integration between measurement and geographical analysis.

For any data layer, there are also problems with the horizontal integration of data across space. The nature of digital data products that are available through national mapping agencies provides a back-cloth against which tactical and strategic analysis may be carried out, and yet the content of digital map series as well as the cartographic conventions used by different national mapping agencies may generate visual differences where national boundaries are traversed. The differences will be more than visual if mapping products are used for distance and other explicitly geographical calculations, for the scale and accuracy with which input data are encoded will influence the recorded measurements that may be made.

Finally, there are problems regarding the different areal entities that are used as the basis to data collection and dissemination between different countries. We have seen in Chapter 2 how incompatibility between the different layers of a geographical database can create errors, overlaps and other analytical difficulties. As regards the creation of pan-European databases, there are clearly additional problems *within* as well as *between* each layer of a database which traverses more than one nation or other administrative system. In the following chapter, we will begin to explore how the richness of census and other information might be unlocked through the development of models of continuous geographical distributions independent of areal zoning schemes. In this way, we will illustrate how problems with the vertical and horizontal integration of geographical data-sets might be tackled and resolved.

Censuses and the modelling of population in GIS

David Martin

4.1 Introduction

Information derived from censuses of population remains one of the most significant sources of data about socio-economic conditions, despite the growth of extensive databases relating to individuals. Consumer databases, however attractive, rarely represent more than 10 per cent of total population, and may be extremely socially biased in their coverage (notwithstanding Openshaw's statement, Chapter 7, this volume, of their usefulness for market analysis: see also Birkin, Chapter 6, this volume). Census information can be of enormous interest to organizations whose business involves them with understanding population characteristics or behaviour. Examples include the retailer who wishes to determine the right store format to suit local tastes in a new area; the health authority manager who needs to predict demand for services within a catchment area or region; and the utility company which wishes to build models of demand for its services in the light of varied social and economic conditions. Geographical information systems (GIS) offer a powerful range of tools for these people, and the increasing availability of large digital data-sets relating to population is expanding this potential at an ever-increasing rate. This chapter reviews this general field, and takes a critical view of the kinds of application which can be developed. My first thoughts on drafting it were to give it the title: 'Population models in GIS: the awful truth'. Although this was abandoned in favour of a less alarming title, the main message is still a cautionary one: there is enormous potential here, but any attempt to represent the socio-economic environment in GIS involves some form of implicit or explicit modelling, and users need to be fully aware of the implications of their actions!

Although there will be situations in which the users have extensive personal databases of their own, perhaps based on customer records or specially commissioned market surveys, there is still frequently a need to relate this information to the characteristics of the background population. National databases such as censuses or population registers generally achieve a degree of coverage and breadth which a non-governmental organization cannot hope to piece together. Such large-scale data-sets are made available at the aggregate level, and a major task involves the correct representation of these data-sets within the computer, and their integration with more specialized data-sets relating to individuals. The late 1980s and early 1990s have seen a massive explosion in the amount and availability of georeferenced information about population, especially using postal geography as the georeferencing mechanism (Raper *et al.* 1992; Martin and Longley, Chapter 2, this volume). Frequently, the organization collecting the data does not conceptualize its database in explicitly geographical terms, but includes geographical references as part of address records which may be used for routine administrative purposes. The expansion of georeferenced data-sets of this kind is generally in keeping with central government policy (e.g. Department of the Environment 1987), but increasingly also demands careful management and integration in order to obtain the maximum returns from investment in data collection and entry (Cassettari 1993). It is in the context of rapidly growing spatial databases that geographical information systems have attained their present significance.

The period since the early 1960s has seen the continual growth and development of GIS, although the massive expansion of the field in commercial terms is a particular phenomenon of the late 1980s and early 1990s. Many early GIS applications concerned the modelling of aspects of the physical environment, both natural and man-made. Examples include applications in soil science (Burrough 1991), and urban planning (Parrott and Stutz 1991). Computer hardware capabilities were also of significance in early systems, providing real constraints to what was possible in some fields. A review of GIS experience in the US (Tomlinson 1987) identified a number of principal application areas, and it is significant that none of these directly involved the use of census or other socio-economic information. This is not to say that no work had been done on the processing of geographically referenced population information within computers: simply that it was relatively late to come under the GIS umbrella. A notable exception to this pattern was the development of the DIME census mapping system in the US for the 1970 census data, discussed below. Since the late 1980s, we have seen increasing use of GIS for socio-economic applications, as discussed more fully in Martin (1991), and it is within this context that our discussion of censuses and population modelling for business use in GIS should be set.

In this chapter, we shall review the sources of information about population in an international context, and focus particularly on population censuses. The UK 1991 Census of Population is taken as a case study, and potential GIS applications are considered in a way which is relevant to business users. The innovative application of GIS to population data is highly dependent upon the nature of the geographical referencing available in the source data, and the way in which this information is conceptualized and modelled within the system. The final part of the chapter sets census applications in the context of contemporary discussion about the representation of population, and reflects on some of the most significant new developments in this area.

4.2 Sources of population information

The collection of large-scale information about population is reliant on either the conduct of censuses, the maintenance of population registers, or attempts to estimate population size by indirect means such as satellite remote sensing (Rhind 1991). Population registers exist in one form or another in most European countries, and have proved particularly successful in Scandinavia, Belgium and Luxembourg. These schemes involve the compulsory registration of residence for all individuals, thus providing a constantly updated central database of basic population counts and age/sex information. In contrast, there is no history of such registers in the United Kingdom or the United States. In order for population registration to be of value, it is necessary to achieve a very high degree of enumeration accuracy. Redfern (1989) observes that accuracy is related to the amount of use which a register receives, as increased use provides more opportunities for updating, and the registers which form part of some European integrated administrative systems may reduce error to as low as 0.3 per cent. National registration statistics are primarily collected for administrative and fiscal purposes (Oberg and Springfeldt 1991), and may fail to portray accurately certain groups such as short-term visitors. Registers generally contain only a few fields of core information such as name, age, sex, and address, but may be used as a sampling frame for surveys or censuses. In Sweden, for example, statistics from the national population registration scheme are supplemented by periodic population censuses, and individual census returns may be linked back to the population register. The registration of individuals living at each address is linked to an advanced system of property georeferencing which allows the rapid production of geographically detailed population information in computer-readable form. Ottoson and Rystedt (1991) illustrate this capability, and describe a GIS application concerned with transportation planning based on detailed local population information. The possibility of building a population registration system was briefly considered, alongside other alternatives, in the UK following the 1991 census, but the stage seems set for another 'conventional' census of population in 2001 (OPCS 1993). In other countries, where neither the operation of a national population registration scheme nor the collection of periodic censuses is practical, some use has been made of satellite remote sensing in order to estimate population totals such as that discussed by Lo (1989) and Rhind (1991). Censuses remain the primary method for the collection of wide-ranging socio-economic information, based on either a large sample, or entire population, and it is on census data that we shall now focus. International differences are also apparent in the form and content of census questions between different countries, with a wider range of topic coverage in countries such as the UK, which have less well developed central registration and administrative systems. Variations are found in the social acceptability of different question topics, with a question regarding income being used successfully in the US, but not considered appropriate in most European censuses (Marsh 1993a). Most business GIS applications will in reality gain a significant part of their socio-economic detail from non-census sources, including for example financial details from existing customers, and commercially available market research data. Census data-sets do however provide an important framework for the georeferencing of base populations, to which these more dynamic data-sets can be used to add texture.

Conducting a census involves the identification of the entire target population, and the ability to conduct checks to ensure that everyone has been reached, and that the information obtained is of acceptable accuracy. A significant amount of prior knowledge is therefore required in order to plan and administer the enumeration process, and much of this information is geographical, involving the production of address lists, identification of enumerators' areas, and construction of zones for the aggregation and publication of data. Census information is necessarily aggregated, because of the need to preserve confidentiality, and this aggregation process has implications for the geography as well as the statistical usefulness of the published data (see Martin and Longley, Chapter 2, this volume). The development of large-scale digital mapping and other data-sets which provide locational information related to population is therefore significant to each stage of census organization. Indeed, in the US and Canada, the need for a geographical framework for the census organization has actually been the driving force behind the creation of major new digital geographical data-sets.

In the US, following experimental mailing in 1960, census forms have been mailed out to most households, and the majority are also mailed back. Enumerators are therefore required to visit only a small proportion of the enumerated addresses, in order to follow up missing forms. The ability to organize the census enumeration in this way is dependent on the existence of an accurate national address list, and geographical information which allows addresses to be assigned to particular enumerators, and aggregated into reporting zones. There is constant pressure for detailed data to be made available for smaller geographical areas. This requirement must be set against the absence of a large-scale national mapping series such as that which exists in the UK. In response to this need, the DIME (Dual Independent Map Encoding) system was developed for the US 1970 Census, initially for metropolitan areas only (Barr 1993a). The system comprised a series of files, which contained records for each street segment, together with grid references for street intersections, and information about city blocks on either side of each street. DIME could be used together with an address coding guide in order to assign address records to streets or blocks, or to list all the addresses within a specified street or block. In addition, the system provided a mechanism for simple street and census mapping, and was a major development in terms of its data structure (Peucker and Chrisman 1975).

The DIME system was extended to cover most urban areas for the 1980 census, and in 1990 was replaced by the new TIGER (Topologically Integrated Geographical Coding and Referencing) system. The TIGER database comprises a hierarchy of files, which include both the DIME-type street level information and additional levels of area cataloguing, which contain definitions of political and statistical areas, both current and historic. It is the intention that TIGER should be maintained as an ongoing national geographical database appropriate for georeferencing of addresses, and for use as a sampling frame for population censuses and surveys. As its name suggests, the topology of street segments, intersections and enclosed areas is a key feature of TIGER's data structure (Broome and Meixler 1990). The 'topology' here describes the structure of the street network in terms of links and nodes, identifying which streets meet at each intersection, and providing the basis for extracting specific city blocks and tracing complex routes. Information about the shapes of street segments has enhanced the usefulness of the data as a base for digital map production. The database provides national coverage without gaps or overlaps, and has formed the basis for many new GIS applications and software products which are not directly related to its original census purpose. US freedom of information legislation has ensured that the TIGER data have been made widely available, and a variety of value-added versions of the national database are available.

In Canada, a comparable system has been developed in which digital boundary files are nationally available which correspond to the geography of the 1991 census. Detailed street network files (previously called area master files) are available for most large urban centres. As in the US, the street network data lend themselves to GIS applications such as route planning, where it is necessary to relate population distribution to other features of the built environment. Again, there is a close link between the census geography and postal codes, and a conversion file exists which allows the assignment of postcoded addresses into census areas, and the provision of corresponding grid references (Statistics Canada 1992). In the UK, as will be seen below, the relationship between census geography and the physical environment of streets and addresses is less clear.

It is of fundamental importance to recognize that there is a sense in which none of these geographical referencing systems provides the 'correct' geographical framework for the handling of population data collected at the time of the census. In all such cases, the boundaries of city blocks or enumerators' areas will be arbitrarily related to the underlying characteristics of the population. In some cases, the census boundaries may be significant socio-economic boundaries, while in others the boundary may split otherwise homogeneous population groups (see the discussion of the modifiable areal unit problem in Martin and Longley, Chapter 2, this volume). The aggregate nature of the published census data makes it impossible to resolve these issues unambiguously. An additional problem concerns the fact that most census geographies fill the entire land area, and it is thus impossible to extract clearly the true geographical distribution of population (which is in reality very sparse in most countries) from the zone-based census data. These are crucia in any attempt to develop GIS applications for use with population data censuses, whose importance is independent of the particular national census with which we are dealing. All representations of population within GIS are simply models of a distribution whose precise characteristics are unknown. If the maximum benefit is to be obtained from such applications, then careful attention must be given to the way in which population is represented within the system. In the following section, the UK census is taken as a more detailed example, illustrating the key points in relation to data publication, geographical referencing and current trends about which international comparisons can be drawn. The UK data also serve as an example for the illustration of the construction of geodemographic indicators as discussed in later sections (Batey and Brown, Chapter 5, this volume; Birkin, Chapter 6, this volume). Many of the issues concerning the geography and content of the UK census are of broader relevance to the interpretation of other national censuses and surveys, and Waters (Chapter 3, this volume) describes the geography of a variety of these sources.

4.3 The 1991 UK Census of Population

The most recent census in the United Kingdom took place on 21 April 1991, the latest in a series of decennial censuses which started in 1801. As in other countries, the censuses of the early 1990s may be considered to be the first to have taken place within an environment of widespread GIS awareness and use. Computer-readable statistical abstracts have been produced from the UK census since 1961, and each census since then has seen successive increases in the range and complexity of the computer-readable information produced. Increases in computing power have made possible the production of more information in a shorter period after data collection, and the computer-readable data are now seen as the principal form of output. A number of different georeferences are available for the 1991 data, and these are important to any attempt to integrate the census data with other forms of geographical information within GIS. There is no comparable set of digital boundary data which links census geography to street networks, as in the US or Canada, nor is there a directly georeferenced national address list which was complete at the time of the census. In this section we shall consider the characteristics of the 1991 census in some detail, and highlight those issues which are particularly relevant to its use in business GIS applications. Major sources of detailed information about the 1991 UK census are Dale and Marsh (1993) and Openshaw (1995); a shorter overview is found in Martin (1993). The UK census is organized, and output data published by the government Census Offices: the Office of Population Censuses and Surveys (OPCS) in England and Wales, the General Register Office (GRO(S)) in Scotland, and the Census Office (Northern Ireland). Technical information is provided in a series of Census User Guides produced by the Census Offices, and in OPCS (1992a).

The Census Offices are concerned only with the production and publication of statistical information, and do not provide any software for the manipulation or analysis of the census data products.

4.3.1 Data collection and processing

The standard 1991 census form in England contained 25 basic questions on 12 pages, covering demographic and economic characteristics, housing, household composition, household spaces and dwellings, and a range of employment and social class characteristics. In addition, there were some country-specific language questions in Northern Ireland, Scotland and Wales. The UK census is considered to be one of the more comprehensive national censuses in terms of its topic coverage. The forms were delivered to each household in the week preceding 21 April by census enumerators, who were temporary employees of the Census Offices during the data collection period. Enumerators called back in the few days immediately following census night, and households which could not be contacted were requested to post their forms back to the Census Offices. Each enumerator had responsibility for an enumeration district (ED), and there were 130 000 EDs in England, Scotland and Wales. An 'average' ED contained about 200 households and 400 individuals. In Scotland, EDs were, on average, slightly smaller. Institutions in which 100 or more people were expected to be present on census night were classified as special enumeration districts. EDs form the basic spatial unit for administration and data output. In England and Wales, EDs nest into wards, which nest into local authority districts, as illustrated in Figure 4.1. The hierarchical organization of enumeration is another common feature between various national censuses. A set of counts known as the Small Area Statistics (SAS) provides information

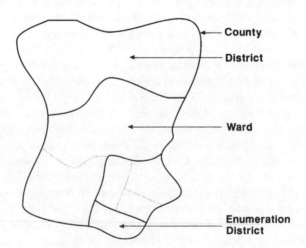

Fig. 4.1 1991 UK Census geography

in around 9000 cells for each ED in England and Wales, or Output Area in Scotland, and this is the most geographically detailed published output.

The information on the forms was then entered into the census database, the answers to each question being coded, and subject to a number of checks for internal consistency. Postcode of address was included in the computer records, but names and addresses were not encoded. Processing of the census questions was divided into 100 per cent and 10 per cent processing, and was performed in two separate stages. For 100 per cent questions, information was encoded from every census form, but for so-called 10 per cent questions, only a 10 per cent stratified sample of forms were processed. The 10 per cent questions are generally those relating to occupation and socio-economic characteristics which are more complex to code. Thus information is collected from the entire population, yet 90 per cent remains uncoded for a subset of the questions. This contrasts with the sample censuses which are used periodically in those countries with population registration systems, in which a similar result is achieved by selecting the required sample, using the population register as the sampling frame. In the US, a more detailed 'long' census form is only sent to a sample of individuals, thus avoiding the collection of data which are not coded. For the first time in the UK, in 1991, characteristics were imputed for wholly absent households, and absent usually resident individuals from enumerated households were imputed in the 100 per cent data.

In all countries, census data are subject to a number of constraints which protect the confidentiality of individual records. The level of protection which is required varies according to culture and political regime. In the UK, the disaggregate information remains confidential for 100 years, and all data output is therefore based on geographical aggregations, using the enumeration district as the basic building block. Published data consist of a number of paper reports relating to specific regions or themes, and a range of computer-readable statistical abstracts. The most widely used of these are the Small Area Statistics (SAS), mentioned above, and the Local Base Statistics (LBS), which provide a fuller range of counts at ward level and above. Additional confidentiality constraints include the restriction of data for any area falling below basic population thresholds of 50 individuals and 16 households (SAS) and 1000 individuals and 320 households (LBS), and the quasi-random modification of counts in the range −1 to +1 (SAS) and −2 to +2 (LBS), to avoid the inadvertent disclosure of information relating to identifiable individuals. A large range of other more specialized computer-readable data products have been produced from the 1991 data, and include the Special Migration Statistics (SMS) and Special Workplace Statistics (SWS), which contain detailed information on flows of migrants and journeys-to-work; and the Longitudinal Study (LS), which comprises a sample of individuals originally drawn from the 1971 census, which were monitored again in 1981 and 1991 (Dale 1993). A new product for 1991 is the Sample of Anonymized Records (SAR), specially commissioned by the Economic and Social Research Council (ESRC). The SAR data-sets contain two samples: a 1 per cent sample of households and a 2 per cent sample of individuals, for which the complete anonymized records are available. In general, records are geographically referenced to the local authority district level (Marsh 1993b). Samples of anonymized records (also referred to as census micro-data) are already well established in the US, Australia and Canada. The research potential of these special abstracts is enormous, but the major use of census data within GIS is likely to involve the detailed small area data-sets and it is on these that we shall concentrate. We are thus concerned with scores and ratios relating to area and to a lesser degree points. In Batey and Brown (Chapter 5, this volume) and Birkin (Chapter 6, this volume), the construction of multivariate indices for these geographical areas, and their integration with individual data, are covered in depth.

4.3.2 Geographically referenced data

The SAS and LBS provide detailed cross-tabulations of responses to the census questions, and are available for various levels of the UK census geography. The SAS are a subset of the 20 000 LBS counts, and are available at the ED level. It should be stressed that the 10 per cent data for small zones may be subject to large sampling errors, and are provided primarily as a building block for the construction of larger areas. It would be highly misleading to import the 10 per cent data into a GIS application and treat the values as though they were 100 per cent data. EDs are designed primarily in order to facilitate the administration of the census itself, and are thus not always ideal geographical areas for analysis. The key requirement of ED boundaries is that they should nest within higher-level administrative boundaries, which are themselves subject to periodic change between censuses. ED boundaries also need to be adjusted in order to standardize estimated enumeration workloads, and to accommodate residential demolition and new development. As a result of all these processes, EDs vary widely in geographical extent, and frequently have highly irregular boundaries which cannot be readily described in terms of the street segment topology used in North America; there is also considerable change in ED boundaries between censuses, making the direct comparison of local statistics difficult. In Scotland, EDs were used only for data collection, and separate Output Areas were designed, which corresponded as far as possible to the EDs for which data were published following the 1981 census; thus the situation in Scotland is more favourable than in the rest of the United Kingdom. Following the 1971 census, SAS were produced for Ordnance Survey grid squares, which have the advantages of regular size and shape, and stability over time. Unfortunately, this product has not been repeated in 1981 or 1991. Good examples of the computer-generated maps which were possible using the grid cell data long before widespread GIS use will be found in the census atlas *People in Britain* (CRU/OPCS/GRO(S) 1980).

National digital boundaries for the 1981 census geography were only created at the ward level, and digitizing of ED boundaries took place locally to a range of standards and with incomplete coverage. For 1991, there are two complete national sets of digital boundaries for EDs and all higher level spatial units, which are available commercially. The two boundary products are ED-Line and ED91 (Dugmore 1992). ED-line incorporates digital ward boundaries created by Ordnance Survey. In addition to the digital boundary information, each ED has been assigned a population-weighted centroid location which is included in the SAS header information. These centroid locations have been determined by eye at the time of ED planning, and are intended to represent the centre of the residential part of each ED. The grid references are given to 100 m in rural areas and 10 m in urban areas.

It was originally intended that the 1991 census geography in England and Wales should be constructed from aggregations of unit postcodes, as was the case in Scotland in both 1981 and 1991. The postcode system is also hierarchical, with unit postcodes (typically 14 addresses) nesting within sectors, districts and areas. Postcodes are increasingly widely used as a referencing system for personal records held in both the public and private sectors (Raper *et al.* 1992). Advantages of the postal geography include the relatively small size of unit postcodes, constant updating, and widespread familiarity among a range of users with the system. In Scotland, digital boundaries have been created for unit postcodes (Clark and Thomas 1990), which allowed enormous flexibility in the creation of the census geography, but in the mid-1980s resources were not available for the definition of boundaries for all unit postcodes in England and Wales, and the proposals for a postcode-based geography were abandoned. As an alternative, it was intended that SAS might be produced for 'pseudo-EDs' based on aggregations of unit postcodes, but these plans were also dropped, although postcode of address was still recorded on the census forms. A directory of EDs and postcodes has been created for England and Wales, which describes each 'partial postcode unit', and is discussed below. In Scotland, index files are available which define the postcode composition of the census Output Areas and higher-level census zones.

A consequence of the continually falling cost of computing power is that there are many more users of the computer-readable abstracts in 1991 than was the case in 1981. In 1981 a single software product, SASPAC (Small Area Statistics Package), was developed in order to facilitate the production of reports and tabulations of the SAS. SASPAC read the data files created by the Census Offices, and offered area selection and the computation of new variables through simple arithmetic operations. The software was widely used in local authorities and universities, running in batch mode on multi-user computers. By the late 1980s, it had become possible to purchase the entire 1981 SAS on CD-ROM with PC software, and many users were conducting relatively complex analysis and mapping of census data using PCs. Two PC-based software products are available for the manipulation of the 1991 SAS and LBS: SASPAC91 and C91. SASPAC91 has been developed from the original software and is available across a range of hardware, with a menu-driven interface on PCs and work-stations. C91 is an entirely new product developed in a local authority context, and provides menu-driven access to the data on PCs. In addition, the data are again available on CD-ROM, packaged together with mapping and retrieval software. For many purposes, it will be desirable to extract subsets of the census data for processing within more specialized statistical software, and this is a straightforward step. As a result of these developments, many more business users have direct access to the small area data, and will have a wide range of software tools for its manipulation (see Cresswell, Chapter 9, this volume). For many of these, geographical analysis of the census data will be an important dimension to their work, and contemporary GIS offer a new environment for such projects. This trend towards increased use and wider access is directly comparable to that in the US, where the TIGER data are available at low cost on CD-ROM, with the exception that the UK Census Offices are charged with a more stringent responsibility for cost recovery. The pricing of the UK data will thus inevitably delay or prohibit some of the small value-added developments which might be seen elsewhere.

4.4 Using census information in a business GIS context

Having considered the sources and form of census data, we can now turn to some of the general principles governing its use in GIS. Census mapping using computers has conventionally involved the use of digital boundary information and zone data such as the UK SAS counts in the construction of shaded area (choropleth) maps. There were many examples of such applications following the censuses of the early 1980s (e.g. Bateman *et al.* 1985), and even earlier examples using the DIME-type information available in the US and Canada. Area-shading has been the most widespread method for census-based population mapping within GIS applications. This typically involves the storage of the digital boundary data in a segment or zone-based vector data structure, and the zone data in a linked data table, which is one of the simplest ways of organizing information within a GIS. The commercial availability of digital boundaries in a variety of widely accepted data formats makes this form of mapping within contemporary GIS a relatively straightforward operation, and a number of simple mapping packages which enable the user to explore data in this form have been produced and marketed specifically in association with census data-sets. Another group of tasks to which GIS have been applied concerns the allocation of geographical objects (such as addresses or postcodes) into census areas or vice versa. An example of such an application would be the allocation of customer address locations into areas for which population statistics are available, in order to assess customer characteristics. In many cases, these list-matching operations could be equally well performed by database software, and do not involve any explicitly geographical processing (Barr 1993b).

Where directories for matching such data do not already exist, GIS provide tools for the allocation of points into areas, or the reallocation of data values between different areal units. For some users, the ability to visually and interactively explore geographical patterns in census data will be a valuable end in its own right. However, not only is this a very low-level use of GIS, but, as has already been stressed, even the very simplest form of shaded area map produced with a mapping package or GIS can provide only a single *model* of the true geography of population, and its various attributes. The results of such applications are therefore heavily dependent on the boundaries used, and are particularly prone to errors in published boundary and directory data. The census data have already been aggregated and presented for a set of zonal units whose boundaries may have little or nothing to do with geographical patterns in the underlying phenomena (Morphet 1993). As cartographers have long recognized, this fact is important if we are seeking to interpret a printed map intelligently, but is of far deeper significance if we wish to exploit the manipulative power of GIS in order to produce more complex analyses. All information within a GIS is in a sense an abstraction or 'model' of aspects of the real world, which incorporates assumptions about the significance of boundaries and data values. It is therefore essential that the user fully understands the implications of the data model which is being used, and these issues are discussed more fully in the following section.

Even before the widespread availability of GIS tools, geographically referenced census information already played an important part in the development of both public policy and commercial strategies. Not only does the basic patterning of demographic and social phenomena provide a general backdrop for policy formulation, but the spatial distribution of population is a fundamental input to many important decision-making processes. This is particularly the case where the geography of population is in some sense significant to the organization, and examples include site selection, catchment area definition, market analysis, financial allocation and political districting. Complex multivariate indicators have been derived in many different sectors which provide summaries of socio-economic variation across space. In the context of resource allocation for health care in the UK, there has been much debate about the use of indicators which combine census data, weighted according to expert knowledge, in order to provide 'deprivation' indicators used in the allocation of finance (Jarman 1983; Clarke and Wilson 1994; Hirschfield 1995). In the field of market analysis, a number of competing neighbourhood classification schemes have been developed which rely on clustering algorithms or principal components analysis in order to identify geographical areas of similar neighbourhood type (Brown 1991). The increased availability of census data in the early 1990s, combined with the existence of more comprehensive in-house databases giving information about patients, customers or survey respondents, seems likely to ensure the development and enhancement of such indicators. Developments may be seen both in a wider range of application fields, and in the incorporation of a greater degree of business information from organizations' own sources, providing customized indicators. The construction of such indicators is covered in more detail elsewhere in the chapters by Batey and Brown (Chapter 5), Openshaw (Chapter 7) and Birkin (Chapter 6) in this volume.

The ready availability of large quantities of georeferenced non-census information about population from surveys and organizations' own records, together with the extensive new census data-sets, with a range of associated geo-codes, have opened up exciting opportunities for the development of new GIS applications. Many of the applications introduced above were originally developed using *ad hoc* computer programs or statistical packages which were not able to provide an integrated framework for the management of such diverse geographical data-sets. GIS, however, provide an environment for the integration of such data, and should (in theory at least) offer the user an extensive range of facilities for making use of almost any of the existing techniques, together with the potential for the development of more sophisticated analyses. Most contemporary GIS do not incorporate the specialist statistical functionality which is required in order to perform (for example) cluster analysis of a set of geographically distributed variables, and these are often useful in a business context to perform neighbourhood classifications. However, they increasingly create either the ability for the sophisticated user to construct his or her own functions using libraries of processing routines, or to readily exchange information between the spatial database in the GIS and separate statistical packages. The GIS environment provides an opportunity to integrate data from different sources, perhaps using different georeferencing systems, and to establish the geographical relationships between these. GIS are uniquely able to provide spatial search and analytical functions which allow the creation and manipulation of new geographical objects, through functions such as interpolation, neighbourhood operators and distance-based calculations.

The quality of the answers obtained from processing of this kind is largely dependent on the structure and quality of the input data, and the way in which they are modelled within the system. For example, the topological structure of the US TIGER data makes them ideally suited to the support of redistricting algorithms, allowing recombination of base units in order to create new geographical objects: objects which are meaningful in terms of the built environment of streets and city blocks. It could be argued that it is only when applications make use of the more complex spatial analytical functions that are now becoming available, that the real power of GIS is being utilized. Business users may have much to gain by use of the right interpolation algorithm for estimating market potential in hitherto unexplored areas, or by incorporating neighbourhood functions which build realistically on the locational uncertainty in their data-sets.

It is this ability of GIS to manipulate and remodel geographical data in order to answer a range of explicitly geographical questions that distinguishes them from computer-assisted cartography and image processing systems. Some examples of the sorts of geographical questions which might be of interest in the context of population data are suggested in Table 4.1. Using these questions as examples, we shall explore the role of the data model in determining the quality of the eventual answer. In the following section, we shall explain more the role of this 'data model', and explore some possible

Table 4.1 Examples of explicitly geographical questions

1. Where has the fastest population growth taken place?
2. What are the characteristics of the people who live within 15 minutes' drive time of this location?
3. How many of the households in this town are classed as socio-economic groups 1 and 2 ?
4. Which postal codes correspond with this area of public-sector housing?

designs. This discussion will illustrate why questions such as those suggested in Table 4.1 are actually very hard to answer unambiguously.

These types of question demand explicitly geographical processing in addition to operations which may be performed on the non-spatial aspects of the database. For example, Question 1 in the table may involve a relatively simple search of the population database in order to determine the magnitude of population change, but the answer must be expressed in geographical terms (perhaps as a shaded area map, or a list of place names). Question 2 requires a higher degree of geographical processing, in which it is first necessary to establish which areas can be reached within 15 minutes' driving time, probably by use of digital road centre-lines and a networking function. A simpler but far less precise solution might be found by using a simple crow-fly distance corresponding to a 15-minute drive, and producing a buffer zone of this width around the location of interest (but see Birkin, Chapter 6, this volume). It is then necessary to establish the population characteristics which apply to these areas, for which a range of different geographical criteria might be used. For example, we might take the population of all the zones through which a main road passes, or all the zones whose centroid falls somewhere within the drive-time limit. Each of these groups arguably lives within 15 minutes' driving time. The third question implies that the town of interest can be clearly isolated as an entity or entities in the database, but there may in fact be many such definitions, including commercial, administrative, functional, political and census boundaries. Any answer to this question requires the identification and retrieval of information about socio-economic groups from one of these specific boundary sets, which may not be the same zones for which the data have been collected and stored. The final question in the table involves the isolation of a geographical area containing public-sector housing, perhaps from census data, and its association with a second set of geographical references, with which there may be no direct correspondence. In summary, it would be possible to arrive at many different answers to each of these questions, even from the same input information. The actual answer achieved will be fundamentally influenced first by the way in which population is stored and represented within the GIS, and secondly by the geographical processing algorithm which is used. In many situations, these factors will be hidden from the end users, unless they choose to delve more deeply into the functioning of their system. A far more detailed and technical explanation of the ways in which GIS are able to tackle problems of this type may be found in Laurini and Thompson (1992).

4.5 Models of population in GIS

It is perhaps most helpful to begin our discussion of models of population within GIS by considering the fact that the information within a GIS is only an abstraction, or 'model' of the real world situation. Figure 4.2 illustrates the process by which data about the real world are collected and stored within the GIS. Whether the collected data concern physical attributes such as soil type, or socio-economic characteristics such as employment status, it is clear that they only capture certain features of the phenomena themselves. The very process of data collection may involve sampling (as in a sample census) or aggregation (as in the collation of unemployment data for census enumeration districts). Geographical location, which is essentially continuous, will be recorded discretely, by means of a grid reference or zone boundary, to a specified level of precision. In both cases, some of the detailed variation in levels and location of the original phenomena will have been lost, and cannot be accurately reconstructed from the data alone. We have also implicitly chosen to record each phenomenon as a particular type of spatial object (in the case of unemployment, as an areal phenomenon). In traditional mapping, certain rules would be followed for the representation of this information on paper (boundary interpolation or area shading), and this would be the end of the process. The enormous power of GIS lies in the fact that we are not limited to any single set of rules for the representation of our data, but are able to construct a data model in which it is possible to alter the way in which operations such as interpolation and area shading are performed. In fact, it is possible to go far beyond this, and remodel the same data as different types of spatial object (from points or areas to a continuous surface, for example). In addition, we have a vast range of tools which allow us to query and manipulate our data model, and to vary the cartographic parameters of scale, symbolization, orientation, etc. The output information may be the result of specific queries in the form of text, tables or map images. Many of the ideas introduced here are explored more fully in Martin (1991, ch. 4).

We shall illustrate the significance of the particular options chosen in relation to the construction of a particular data model. In reality the population consists of a large number of individuals scattered across geographical space, each of whom exhibits various socio-economic characteristics, such as age, sex and employment status. These individuals move through geographical space in daily, weekly and longer-term cycles, but their 'location' is usually thought of as their place of residence, perhaps recorded as an address or postal code. Any attempt to assign geographical meaning to population distributions usually requires the recording of information about many individuals, and its organization into some form of geographical framework.

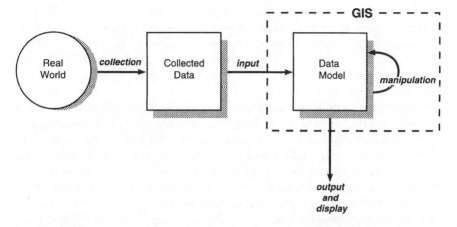

Fig. 4.2 The place of the data model in geographical data processing

We have already considered a number of examples of data collection of this kind: population registers, customer lists held by companies, and particularly census information. Each contains a number of variables describing the characteristics of individuals, and is (or may be) geographically referenced by use of postal or census geographies, which allow us to attach grid references with varying degrees of precision. Unfortunately the recording mechanisms often get in the way of the most interesting geographical information, such that large numbers of diverse individuals are grouped together within a zone whose boundaries are known, but which provides us with no information about internal variation. Alternatively, a postal address may only allow us to assign a grid reference at a relatively low resolution, so that we are unable to determine precisely to which of a number of administrative areas a particular individual belongs.

Thus we see that population, which is in reality an unstable pattern of individuals scattered across geographical space, is most frequently represented by arbitrary aggregations of data into reporting zones. When these data are held within a GIS, they are unable to offer very satisfactory answers to questions such as those in Table 4.1, regardless of the accuracy (or otherwise) of the collected data. Although most large GIS will allow the user to approach a solution, the answers will remain 'fuzzy', with a variety of areas of uncertainty. It is the existence of these which led Openshaw (1989a) to suggest that such uncertainties should be made explicit when answering questions of this kind. The area of fastest growing population (Table 4.1) is likely to be a new housing development which is a subdivision of an existing zone; we cannot tell what proportion of each zone's population are close to the road network; the town may not be a neat aggregation of any group of zones, and the public-sector housing area is likely to be georeferenced using a zoning scheme other than postal areas. These geographical problems cannot be neatly resolved by any degree of boundary topology, and estimation of zone values from those of another incompatible set remains problematic. We would only be able to provide precise answers to these questions if we had information about the location of each individual and his or her characteristics. So long as we do not have such data, we need to represent the aggregate data in the GIS in a way which represents the significant geographical properties of the population distribution as closely as possible.

Fotheringham and Rogerson (1993) identify eight 'impediments' to spatial analysis in GIS, of which at least four are relevant to the handling of population data discussed here: the modifiable areal unit problem, boundary problems, spatial interpolation, and aggregate versus disaggregate models. The modifiable areal unit problem in particular (Openshaw 1984a, 1984b; see also Martin and Longley, Chapter 2, this volume) is fundamental to many of the difficulties of using census data aggregated to reporting zones. These are what we might describe as data model problems. It is argued here that the route to effective GIS applications in the realm of census and population data lies in the realm of data models, and the nature of the input data-sets. As an interim solution, it is possible to improve the ability of the GIS to answer geographical questions by using a more appropriate geographical model, but the true solution lies in more appropriate geographical data!

Given what we now recognize about the true geographical properties of population, it would be an improvement if we could model it as an object which more closely represents the reality. Various researchers over time have argued for the representation of population as a continuous surface (Nordbeck and Rystedt 1970; Tobler 1979). The properties of surface models of population density vary according to the size of the reference area across which density is calculated. Thus surfaces which express population per $100 \, \text{km}^2$ will pick up only the broadest variations in the settlement pattern across a country or region, while a surface expressing population per $1 \, \text{m}^2$ would effectively be a dot map. A surface model constructed from the most detailed small area census data can provide an effective compromise, giving a clear reconstruction of the settlement pattern, and reconstructing the pattern of 'real' geographical objects such as neighbourhoods and isolated settlements. In a recent review of the problems of population data interpolation between incompatible georeferencing systems, Goodchild *et al.* (1993) recognize the need to estimate one or more underlying density surfaces, in order to achieve the desired result. Different researchers have tried a variety of methods for the construction and storage of such socio-economic surface models within GIS (e.g. Sadler and Barnsley 1990), but each has achieved a data model which recaptures significant geographical properties that are lost in the conventional choropleth model. An alternative solution to the problem of representation of population within GIS is to attempt to achieve a better reconstruction of the underlying scatter of points which is the 'actual' population distribution. The potential for such data-sets has been considerably advanced by the creation of detailed georeferenced address lists, such as those compiled for the administration of the US census, or the new Address Point data being produced by Ordnance Survey in the UK (see Plate 2.1). There is an enduring difficulty with such data, however, in that it is politically unacceptable to link individual records from a census with precise locations of this type, and so all analysis must still be carried out at an aggregate level. Individual address referencing may prove a realistic option for organizations who already hold large personal databases, and are able to link these via address codes. In the remainder of this chapter, we shall illustrate further the construction and use of both surface models and hybrid data-sets based around more detailed address referencing.

4.5.1 Surface models

We have already noted that most census-type data are provided to the GIS-user already aggregated, and georeferenced by a set of zone boundaries, centroids or a list of associated points. The modelling of these entities within the GIS fails to provide satisfactory answers to geographical questions such as those listed in Table 4.1. The GIS environment, however, offers the potential for the remodelling of these data into surface representations, which can be held as a complementary view of the population, and will enhance the ability to answer questions which explicitly involve geographical analysis.

The approach to surface construction outlined here has been described more fully in Martin (1989) and Bracken and Martin (1989), but it serves to illustrate the more important characteristics of surface-based approaches to the representation of population data. The model has been developed in the context of UK 1981 and 1991 census data, but is suited to any situation where the zone-level population data can be referenced by a series of population-weighted centroids, or other reference points which summarize the population distribution (such as the 1991 UK ED/postcode directory described above). In this discussion, these will be referred to simply as 'data points'. Initial applications to US census data have also provided satisfactory models. The approach adopted is to take the counts (of population, households, unemployed persons, etc.) which are associated with each data point, and to redistribute them into the cells of a regular grid, which will form the output model. In the UK census applications, a cell size of 200 m has typically been used. The surface generation algorithm visits each data point in turn, and a neighbourhood function examines the distance between this point and its neighbours, in order to establish the areal extent of local zones. In rural areas with large zones, data points will be widely dispersed, while in urban areas there will be a strong degree of clustering. A distance-decay function is then used to weight surrounding grid cells according to the likelihood of their containing some of the population which is associated with the current data point. This function effectively spreads the count associated with each data point into the surrounding area. There will be a high level of dispersion in sparsely populated rural areas, but in densely populated urban areas the entire count may be contained within the cell containing the data point. This process is then repeated for every point in the input data-set, until the entire population at the data points has been redistributed into the grid. Many urban cells will receive population from a number of different data points, but the majority of rural cells will remain unpopulated, thus reconstructing the underlying settlement pattern, which is inherent in the data but is not explicitly recorded in the zonal system used for data collection. Two examples of the resulting surface models are shown in Plates 4.1 and 4.2.

Plate 4.1 shows a surface model of percentage unemployment for the London region derived from 1991 census data. This particular image has been created by overlaying separate models of unemployment and base population. The broad pattern of higher unemployment in the inner areas is apparent, as are individual outlying settlements and neighbourhoods with particularly high unemployment at the time of the census. Although presenting an impressive visualization of the pattern of unemployment in the region, a more important characteristic of the data model is its potential as a database on which extensive spatial analyses may be run. For example, each discrete settlement may be automatically identified and treated as a separate spatial object. Plate 4.2 shows the same region, but illustrates a characteristic of the census itself, namely imputation levels. In certain circumstances, where no return was received from a household or individual, their characteristics were imputed by comparison with neighbouring respondents. The model clearly reveals the different character of the inner and outer suburbs, indicating the greater caution with which the census data for these areas should be treated. Again, the greatest potential for a layer such as this is its use within more complex analyses where some measure of confidence in the census data is required.

A data model of this type offers a number of significant advantages over the conventional zone-based model. In a business context, the ability to perform complex geographical queries without incurring either the financial cost or heavy computing requirements of boundary data-sets will be particularly important. The geography of settlements and neighbourhoods is reconstructed, and unpopulated regions are preserved. Within each settlement, all population totals are preserved. The absence of boundary data makes the model highly compact, and amenable to manipulation within a range of raster GIS modelling packages. Thus the creation of new derived variables, including indices of change over time, may be performed without the further need for areal interpolation to overcome the problem of data referenced to incompatible zonal systems. The accessible nature of the data model also makes possible the development of more complex analyses which can reference population information directly at any location. For example, the cell structure might be used as the basis for a spatial interaction model (see Clarke and Clarke, Chapter 10, this volume), or as a background population for expected rates when considering incidence of rare medical conditions. If we refer back briefly to the questions raised in Table 4.1, we see that the location of population growth may be identified directly, rather than by association with one or more extensive (and largely unpopulated) zones. The drive time information required for Question 2 may be directly related to the grid, allowing us to assess the numbers of people at successive distances from the road network, overcoming uncertainty about the road and population locations. The town in Question 3 is now an analytically identifiable spatial object, comprising a number of contiguous populated cells, whose characteristics may be evaluated, and the postcode locations in Question 4 may be inferred directly from the cell locations, again without the need for intervening directories or point-in-polygon processing.

An important point to recognize here is that the output surface model is affected by the choice of data point locations, cell size and distribution function used. Thus the final model is one position from an almost infinite range of possible outcomes. This should not be seen as a weakness; indeed it might be suggested that one of the strengths of the approach is that it makes explicit our choices about data, resolution and method. A conventional zone-based representation is equally limited by boundary location, zone size and manipulation algorithms, yet there is a widespread tendency to see such models as the single 'correct' representation. These surface representations should be seen as a complementary method for the representation and analysis of population-based data, which overcome a number of the commonly recognized problems associated with zonal data models. The surfaces are efficient to manipulate, and avoid the need for costly (in some countries) digital boundary data. Many conventional analyses such as neighbourhood classification which are applied to zonal data may be reproduced using cartographic modelling concepts on the surface layers. For many business applications, where some form of resource allocation in relation to population is required, such models offer a powerful additional tool in the analyst's armoury, and a number of alternative surface construction methods are available, according to the format of the local data.

4.5.2 Hybrid models

A second area in which the modelling of population data within GIS might be improved in the short term lies in the area of what might be called hybrid models. In the absence of individual-level data, there is still much that can be achieved by the integration of existing data-sets in order to achieve more spatial detail or additional statistical information. A frequently cited area of potential for GIS is that of data integration. The existence of a number of different georeferenced data-sets with relevance to population data raises the possibility of achieving a more detailed and realistic representation of population by careful combination of locational and attribute information from different sources.

One of the most significant aspects of the recent growth of georeferenced data-sets has been the increasing importance of address and postal geographies. Many countries now have highly developed systems of postal codes which were originally designed in order to speed the sorting and delivery of mail, but which have also become *de facto* spatial referencing systems, with a greater level of resolution than that available from census sources. The postcode system in the UK is described in detail in Raper *et al.* (1992), and Marx (1990) stresses the continued importance in the US of the address lists associated with the TIGER structure as a basis both for address-matching and the development of new GIS applications. The UK does not yet have complete national address georeferencing, but Ordnance Survey's new Address Point product promises complete coverage by the end of 1995 (Ordnance Survey 1993). This will provide a unique property reference, including a grid reference, for every postal address which appears in the list maintained by Royal Mail. Gatrell (1989) evaluated an earlier commercial data-set which provided a similar resolution but failed to achieve national coverage, illustrating the enormous power of such a detailed point data-set. For most situations such data-sets may actually provide 'data overload', but many useful products can be simply derived from them. In addition to these national georeferencing systems based on postal geographies, various public bodies hold property registers: for example the council tax registers compiled for new local taxation arrangements in the UK (Martin and Longley, Chapter 2, this volume).

Where these detailed postcode and address lists contain grid references for smaller units than those for which census data are reported, there may be scope for providing additional information about the within-zone distribution of the census counts. Association between the list data and the census will be possible either by the use of directory files where these exist, or by geographical manipulation within a GIS. In the US, the Census Bureau's address list allows the assignment of addresses to blocks which are identifiable in the TIGER data. In the UK, the Postcode Address File (PAF) provides a link between all addresses and a postcode, and a further directory indicates the census enumeration districts within which these postcodes fall (Martin 1992; OPCS 1992b). In the absence of such directories, linkages between census zones and lower-level georeferencing may be established by point-in-polygon processing (for point data) or polygon intersection (for smaller sets of zones). Taking the directory of EDs and postcodes as an example, we can see how we might improve the locational information associated with our census data.

From the 1991 UK Census, a directory of enumeration districts and postcodes has been produced which provides an entirely new geography by the creation of a national mosaic of part postcode units (PPUs), and pseudo-EDs (PEDs). The directory structure is illustrated in Figure 4.3. Each record represents a PPU, which is the unique area of intersection between a postcode and ED. Field (a) contains the ED code, and (c) the postcode defining the PPU. Field (d) indicates the ED to which we would assign each postcode on the basis of the greatest overlap (of population), and this is the PED. Fields (e) and (f) contain the grid reference for each postcode, although at present there are no digital boundaries for these new zones. Field (g) indicates the number of households in the PPU, while (b) and (h) are additional indicators concerning the derivation of the data. The proportions in which the PPUs belong to each unit postcode and enumeration district are thus known, and a variety of standard techniques in GIS permit the generation of boundaries for these areas, such as the calculation of new centroids by averaging postcode and ED

(a)	(b)	(c)	(d)	(e)	(f)	(g)	(h)
ZYFA01	0	RO318DS	ZYFA02	44820	09580	005	000
ZYFA01	0	RO318DR	ZYFA01	44830	09610	009	000
ZYFA01	0	RO318DP	ZYFA01	44820	09610	023	000
ZYFA01	0	RO318DW	ZYFA02	44830	09610	013	000

Fig. 4.3   Example of a hybrid data-set: the UK ED/postcode directory

centroids, and the creation of Thiessen polygons (Boots 1986) around these points. Alternatively, these new data points could form the starting point for generation of a higher-resolution surface model than that outlined above. Such a directory is not a simple look-up table owing to the complexity of the overlap between the census and postal geographies, but it offers much more information about the location of households (and by implication the people who live in them) than can be provided by the digital ED boundaries alone.

In any context where we are able to link the census data to a more detailed geographical referencing system, it becomes possible to take the information which is reported for each zone, and to re-express it as a property of the more detailed system. This will not provide any solution to the aggregate nature of the census counts, but will provide important additional information concerning the location of individuals and households within the zones. In the language of our previous discussion, it has become possible to produce a data model where population is treated as a large scatter of data points with local information associated with them. In many cases this will provide better answers to those explicitly geographical questions (drive time, postcode–census matching, location of development etc.) which are the very area in which GIS have most to offer.

4.6 Conclusion

Population data are vital to the planning and marketing of many different forms of business activity. Although organizations increasingly hold their own individual-level databases relating to customers or clients, there are situations in which this information cannot replace the national coverage and breadth of topics which are collected periodically by central government. There are a number of different schemes in existence for the collection of population data, and the quality and style of this information varies between countries. Although continuous population registration is primarily a feature of the European countries, national censuses of population in some form or other are almost universal. In this chapter we have concentrated on data from censuses, and have considered the different ways in which GIS may be used to handle information relating to population. Theoretically, at least, GIS should be able to offer most in the context of analyses and queries which involve explicitly geographical processing of population data. More routine tasks such as the matching of address lists, management of census data or the simple mapping of counts into zones may be performed within most GIS software, but are not in themselves a requirement for GIS. Many such operations will be performed at least as effectively by other types of (less expensive) software.

Internationally, there has been a massive expansion in interest in population-related applications of GIS at the same time as the new censuses of the early 1990s. When we come to consider the way in which GIS conventionally handle population information, we find that there is a widespread adoption of traditional cartographic conventions, which require the association of census data with a set of zone boundaries within which the data are collected and reported. In some countries, such as the US and Canada, the boundary products provide important sources of street network and other general reference information. In the UK, the census boundaries are a unique product, not necessarily spatially coincident with other significant boundary sets. Unfortunately, in both cases, the relationship between the boundaries of census-data reporting zones and the underlying characteristics of the population (spending power, political opinion, social class) is unknown. When we attempt to use GIS to ask explicitly spatial questions, in which the calculation of concepts such as distance, connectivity or adjacency of population (drive time or neighbourhood analysis), the representation of population which we hold is frequently unable to provide very meaningful answers (see Birkin, Chapter 6, this volume).

This awareness that the reporting units do not provide the best geographical 'model' for population has prompted various researchers to explore alternative ways of modelling such data within GIS. Two possibilities have been explored in some detail here. The first of these is the remodelling of census data in order to represent the population and its characteristics as a series of surfaces. An alternative approach is to make use of the increasing number of georeferenced data-sets, which provide more spatial detail than that available in the census data, and to explore routes by which we may associate the census data with the higher resolution of these other data-sets. The data models resulting from such processes offer more realistic answers to many of those questions to which the geography of population is critical, and the resulting databases are frequently more efficient to store and manipulate, and cheaper to construct, than conventional zonal systems.

It is not suggested that the many mapping systems in use with recent census data and zone boundaries are not of value. Indeed it is a major benefit of the availability of GIS that so many business users are thinking spatially about their organizations, distribution systems and pattern of demand. However, the user wishing to obtain maximum benefit from geographically referenced population information should carefully consider whether one of the alternative 'models' of population would not provide a more appropriate, complementary view of their data. The tools to perform such work are available within existing GIS software, but users need to take time to understand the implications of the way that their systems work. All GIS impose some form of model on the input information, and no such model is 'neutral': it will always affect the quality of results. The future in this area looks bright. Each year sees new data products and improved georeferencing, such that our broader approach to data modelling may soon by supplemented by more appropriately referenced population data.

Part Two

Geodemographics

The contributors to Part One have developed an exposition of the problems posed by areal entities in the overlay and analysis of diverse geographical data sources. This is a prerequisite to analysis of geographical problems. The focus of the first two contributions to Part Two is upon the combination of census and other external data to form indicators which bear an identifiable correspondence with consumption patterns. These are known as geodemographic indicators. Chapter 7 takes a wider view of changes in information technology which have implications for business and service GIS. More generally, the theme of the three contributions is that by adding value to customer databases, GIS helps businesses and services focus upon the customer's expenditure/consumption patterns in relation to his/her neighbourhood characteristics (see also Grimshaw 1994): it is only then that businesses and services can obtain a complete assessment of their performance in the market.

Until quite recently, the development and application of traditional (census-based) geodemographic systems was taken to be synonymous with the use of GIS in business and service planning. In Chapter 5, Peter Batey and Peter Brown trace the origins of GIS-based research in this tradition to the long-established interests of academic geographers in the social and economic make-up of urban areas. The innovation of computing and quantitative techniques into geography led to the computation of indicators of the socio-economic attributes of small areas, and early geodemographic indicators essentially comprised an extension of this approach. Batey and Brown go on to describe how applied geodemographic analysis was initially technology-led but how, with accumulated experience and demonstrable proof that it worked, it has led to programmes of refinement and redevelopment of earlier systems. Their case study is the development of the Super Profiles geodemographic system, which originally arose from a research-council funded project in academia. This raises a different theme to this book, namely the relationship between the academic research base and applied 'near market' research: this theme is pursued in greater depth using other examples in Part Four.

Traditional geodemographic indicators have been based upon census variables alone, but there are problems in over-reliance upon census sources – for example, censuses are typically only collected every ten years, they are restricted in scope with regards to consumption patterns, they provide only indirect indicators of consumption habits, and (in Europe at least) they are notoriously lacking in detailed indicators of income. As a consequence there has recently been a movement towards integration of traditional census-based geodemographics with other private/commercial data-sets, in order to 'freshen up' information and make it more relevant to particular customer purchasing and service-utilization patterns. This development is raised in Chapter 5 by Batey and Brown as a natural extension of Super Profiles, and is reviewed more fully as the 'Lifestyles' concept by Mark Birkin in Chapter 6. As Birkin points out, there is an emergent tension inherent in these developments: on the one hand census-based geodemographics represents a tried and tested technique which emerged from the research base over a thirty-year period; on the other, it is now being augmented by data fusion and other more speculative technologies which have greater intuitive and theoretical appeal, but which, for all this, simply may not work as well in 'real world' situations. Lifestyles and psychographics are two technological developments that are actually being used and promoted at the present time.

Lifestyles and psychographics are, however, just two of a range of technological developments the applicability and reliability of which are under development within the academic research base. The likely uptake of these technologies is a theme picked up by an avowed protagonist of applications-led research, Stan Openshaw, in Chapter 7. Openshaw reviews a range of developing technologies which offer the prospect of enhanced applied spatial analysis. However, his contribution sounds a number of pessimistic notes, in that he does not see the academic research environment as conducive to worthwhile applied spatial analytical work. Moreover, he notes that although problems of spatial analysis are very much the home territory of the discipline of geography, geographers have no absolute proprietorial rights in this area. This issue of the degree to which academic geography is able to generate business solutions other than in a serendipitous way is picked up again in Part Four.

5

From human ecology to customer targeting: the evolution of geodemographics

Peter Batey and Peter Brown

5.1 Introduction

To some of those working in the marketing field, it might seem that the current concern for building small area census-based classification systems is a very recent phenomenon, linked directly to the enhanced data handling capacity of computers and to the widespread adoption of more rigorous customer targeting systems, primarily since the mid-1980s. In fact, however, the field we now refer to as geodemographics has a much longer history and reflects a number of key developments, not in marketing, but in urban research, notably in the United Kingdom and in North America.

Interest among urban researchers in classification systems has two main dimensions, one largely academic and directed towards theory construction, and the other more practical and action-oriented. In their academic research, urban sociologists and geographers have long been concerned to establish a series of general principles about the internal spatial and social structure of cities. From the 1920s onwards, they sought to confirm these concepts by analysing detailed social, demographic and economic data for selected cities. The focus of this work was on unravelling the relationships between a large number of urban characteristics, rather than on studying any individual characteristic in detail. This work ultimately led to the adoption of multivariate statistical analysis, which generated ecological correlations from small area census data. In many instances the end product of the analysis was an area typology which summarized the most significant features of urban structure.

Urban planners have viewed small area classification systems from a different perspective. They too have been concerned to distil the main sources of social and economic variation within cities, but usually this work has had some practical purpose directly related to policy formulation. In many cases the aim has been to develop a consistent and systematic approach to resource allocation, involving the definition of priority areas to receive favoured treatment. Such areas may be defined in relation to particular policy sectors, such as education, housing or health, or in a more general sense, as in the case of designated 'inner city areas'. The area classification here serves as a composite measure of need and is constructed using census variables, sometimes supplemented by other sources of data.

The pragmatic approach of the marketing analyst has much in common with that of the urban planner. In both cases an area classification system is required to provide up-to-date information that is actionable, and the test of a good system is whether it works in practice. It is perhaps not surprising, therefore, that classifications generated originally for use in public policy-making ultimately found their way into the private sector.

This chapter traces the evolution of geodemographics. In the next section, early research on area classifications by sociologists and geographers working in North America and in the UK is examined. Subsequent experience of developing area classification systems in local government is also reviewed. In Section 5.3 attention shifts towards proprietary geodemographic systems and their use in customer targeting. The growth of the geodemographics industry in the 1980s is charted and the key features of the main geodemographic products are described. Section 5.4 contains a discussion of the design considerations affecting the most recent geodemographic systems. It also provides examples of the targeting potential of such systems. In Section 5.5 conclusions are drawn.

5.2 Precursors of geodemographics

5.2.1 American studies of urban spatial structure

Much of the most influential early work in identifying urban spatial structure was carried out for specific American cities. In the 1920s, the Chicago urban sociologists Park and Burgess (Park *et al.* 1925) were prominent in using empirical urban research to develop and test concepts about the form, structure and processes of development operating within cities. Park's work on defining 'natural areas' in cities – 'geographical units distinguished both by physical individuality and by the social, economic and cultural characteristics of the population' (Gittus 1964, p. 6) – typified work in a field which subsequently became known as human ecology (Theodorsen 1961).

Early attempts at 'within-city' classification, particularly those which involved the definition of natural areas, generally lacked methodological rigour. It was not clear how the various classification criteria (social, housing, ethnicity, etc.) were combined, nor was it evident as to which classification method was used. Despite these shortcomings, however, natural areas once defined remained in use as a summary device for reporting census and local statistics. Rees (1972) quotes the example of the *Local Community Factbook* for the Chicago Metropolitan Area, which in 1960 was still using a citywide application of the natural area concept in which 75 community areas defined thirty years earlier were employed as basic statistical units. Such areas had been classified according to a vaguely specified combination of historical, social, physical, commercial and transportation criteria (Kitagawa and Taeuber 1963).

An undoubted stimulus to research in human ecology was the availability, for an increasing number of cities in the United States, of tabulations of data for census tracts, each with a population of about 4,000. Census tracts had been introduced in 1910 when the US Bureau of the Census agreed to prepare tabulations of such areas of New York, Baltimore, Boston, Chicago, Cleveland, Philadelphia, Pittsburgh and St Louis (Robson 1969). Later decennial censuses added to the number of tracted cities so that by 1960 there were as many as 180 tracted areas. Local advisory committees helped in the definition of tracts and where possible boundaries were drawn to follow permanent, recognizable lines and to contain people of similar racial and economic status and areas of similar housing.

Notable among the studies that made extensive use of census tract data were those of Shevky and Williams (1949) and Shevky and Bell (1955) for Los Angeles and San Francisco. They classified the census tracts of these cities into a number of classes that, because of their geographical proximity, were called social areas. The form of analysis was referred to as 'social area analysis' and centred around three theoretical constructs: economic status, family status and ethnic status. Shevky and his co-workers proposed three indices, one per construct, made up of from one to three census variables, to measure the status of census tract population on scales of economic, family and ethnic status, and to enable tracts to be classified into social areas on the basis of their scores on the indices (Berry and Horton 1970, p. 314). Social area analysis thus used classification criteria unique to each particular case study, which meant that the original analysis was incapable of being replicated in other cities by other research workers. Rees (1972) includes a comprehensive bibliography of studies carried out in the United States and elsewhere following the principles set out by Shevky, Williams and Bell.

Social area analysis was used to perform a variety of functions: to delineate socially homogeneous sub-areas within the city; to compare the distribution of such areas at two or more points in time; to compare the social areas in two or more places; and to provide a sampling framework to enable other types of research to be undertaken, particularly for the design and execution of behavioural field studies (Rees 1972, p. 275).

In its original form social area analysis was severely criticized on two counts: first in terms of its theoretical basis (the theory underlying the constructs); and secondly for empirical reasons (the method of measuring the constructs) (Hawley and Duncan 1957). Efforts were made subsequently to test the correctness of the census variables used to measure the constructs (Bell 1955), by employing factor analysis. This work had some success when applied to Los Angeles and San Francisco, but extension to a wider range of cities revealed the shortcomings of the original choice of census variables (van Arsdol *et al.* 1958). It led to the inclusion of a wider range of socio-economic census variables and to the adoption of factor analysis (or the related technique of principal component analysis) as a standard method for identifying the underlying dimensions of urban social and spatial structure. This development of social area analysis became known as factorial ecology and was widely used by quantitative geographers in the 1960s and 1970s, not only in the United States but also in a range of cities throughout the world (Rees 1972; Berry and Horton 1970). Factorial ecology generally led to the production of maps and cross-sections using factor scores for each of the main factors. In this way it was possible to summarize the main features of spatial variation in socio-economic and demographic characteristics.

In some instances, the scores from two factors were used to cross-classify census tracts. Rees's study of Chicago (Berry and Horton 1970), for example, employs a simple graphical technique to categorize areas according to the economic status and family status of their residents. It was uncommon at this time to proceed one step further and use cluster analysis to create a multivariate classification of social areas. One exception, Tryon's (1955) study of the San Francisco Bay Area, was of limited value because of the imprecise way in which cluster analysis was used. Other researchers found it difficult to reproduce the results that Tryon had obtained (Robson 1969, p. 51).

5.2.2 Small area classifications in the UK

In Britain, early studies of urban spatial structure were hampered by the almost complete absence of small area census data. For many years the smallest units for which published census data were available were the ward and civil parish and even for these the range of information was small. Gittus (1964) describes attempts made in 1951 to define zones within the major conurbations that were relatively uniform in relation to the siting of industry and commerce, the rate of population change, and the age and type of housing. Within zones, distinctive areas, both natural and planned, were recognized and their boundaries determined on 'purely local considerations' (Gittus 1964, p. 9). These divisions and subdivisions were intended to provide a more rational basis for presenting social data than that offered by administrative boundaries. However, it proved difficult to achieve consistency from one conurbation to another and little use was made of the zones in comparative studies.

A more productive initiative was the establishment of the Inter-University Census Tract Committee. This committee, formed in Oxford in 1955, was originally intended to consider the definition of census tracts similar to those used in the United States. The city of Oxford served as the prototype for British census tracts and some 48 tracts were delineated with an average population of 2645 (Robson 1969, p. 44). Although these census tracts were smaller than their American equivalents, they were nevertheless fairly large aggregates, likely to exhibit a high degree of internal heterogeneity. One possible advantage compared with other geographical units was that they were more certain of retaining their boundaries over time, allowing comparisons to be made.

In practice, the British Registrar General's Department decided that, instead of adopting census tracts, it would make data available by enumeration district. Such units were considerably smaller, containing on average less than 1000 people. Data on this scale were purchased for most of the conurbations included in the 1951 scheme and for a number of smaller administrative areas. Members of the Inter-University Committee continued to meet and began to develop a series of comparative studies of urban structure. Gittus (1964) used 1951 small area census data for Merseyside and South-East Lancashire in an experimental project, applying correlation analysis and principal component analysis to a set of 27 census variables for each conurbation. This preliminary work paved the way for a number of studies using 1961 census enumeration district data, including Gittus's study of South Hampshire and Merseyside (Gittus 1963–64), Robson's study of Sunderland (Robson 1969) and the Centre for Urban Studies (University College London) Third Survey of London Life and Labour (Norman 1969; Howard 1969).

The work of the Centre for Urban Studies was probably the most important in terms of the development of method, scale of study and influence upon other research. It was able to build upon a pioneering inter-urban study of British towns. This study, carried out by two researchers at the Centre, Moser and Scott, drew upon 60 variables for 157 towns of more than 50 000 population in England and Wales and classified them into 14 groups, using principal component analysis and a graphical plot of scores from the first two components (Moser and Scott 1961). The study was based on a combination of 1951 census data and other sources of social and health data. Some of the data were cross-sectional, while other variables were intended to measure change over time.

In the Third Study of London Life and Labour, the Centre research team experimented with 1961 census data at the ward and enumeration district levels (Norman 1969). In the case of Inner London they produced a six-fold classification of enumeration districts using principal component analysis and a least-squares cluster analysis method (Howard 1969). Figure 5.1 shows an example of the output from this classification for the area of Camden. Two other features of this work are of interest: the naming of clusters ('Upper Class', 'Bed-Sitter', etc.) and the attempt, in Norman's paper, to develop pen pictures summarizing the main census characteristics of each cluster.

The late 1960s saw an increasing amount of interest among local authorities in small area census classifications. One of the earliest of these exercises was a classification of the 32 London Boroughs undertaken by Frances Kelly of the Greater London Council Research and Intelligence Unit in 1969 (Kelly 1969). Kelly's paper acknowledges the advice given on methodology by Ruth Glass of the Centre for Urban Studies and is important in terms of its emphasis on the choice of available cluster analysis methods. Her study made use of 1966 10 per cent sample census data and was restricted to local authority areas in the first instance. Later, unreported work extended the analysis to wards, but the use of

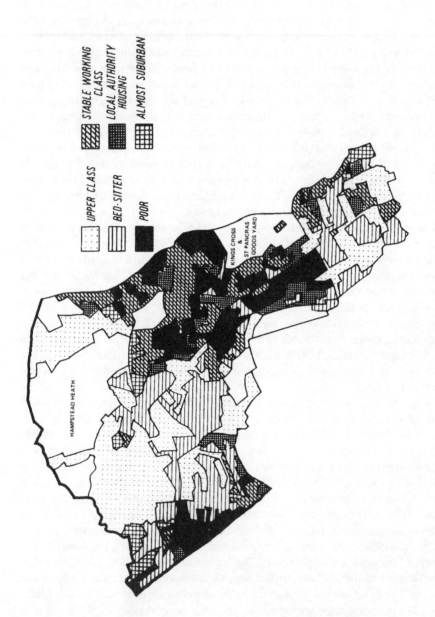

Fig. 5.1 Six types of enumeration district in Camden Census, 1961 (source: Norman 1969)

enumeration districts was specifically ruled out because of sampling and enumeration errors (Kelly 1969, p. 18).

A second paper (Kelly 1971) describes the results of a series of London Borough classifications, six of which employed highly specific classification variables: housing conditions; tenure; socio-economic characteristics; population and household structure; birthplace characteristics; and characteristics of the young and transient population. These topic-based classifications were each produced by the use of principal component analysis followed by hierarchical cluster analysis and none employed more than ten classification variables. A seventh classification was designed to be general purpose and was based on eight key variables that had featured in the topic classifications. Kelly's London Borough classifications were to some degree intended as a demonstration of what could be achieved by the use of multivariate analysis. Her reports were written in relatively non-technical language to promote further use of the methods and included the raw census data to assist in this respect. However, it is not clear whether the classifications she describes had any other immediate purpose.

A second, pioneering local authority study was carried out by Liverpool City Council in 1969 and is particularly interesting because it combined census data with social data drawn from a range of other sources. The Liverpool Social Malaise Study (Liverpool City Council 1969), as it came to be known, was intended to identify areas in the city which had concentrations of social problems. The study was expected to help guide the allocation of social service resources, in the wake of the Seebohm Report in 1968, and the establishment of community development programmes. Unusually for a British local authority, the study team was advised in its early stages by an American social scientist, a Professor Dunham from the University of Michigan (Liverpool City Council 1969, p. 1). The Social Malaise Study employed a variety of statistical methods, including correlation analysis, elementary linkage analysis and principal component analysis. It broke new ground by first carrying out an audit of the social statistics collected routinely by each city council department and then assembling this information at ward level for the whole city. Supplementary information from the 1966 sample census and from the council's slum clearance programme was added to this data-set. The study sought to identify areas within the city where there was a high incidence of social malaise, equating this with a need for extra physical and social resources. It also examined the ecological correlation between the social malaise variables and the 1966 census variables on a citywide basis, so that certain key factors (e.g. possession orders) could be isolated to predict the occurrence of other forms of social malaise. A further aim was to seek to relate 'social' problems to housing and economic factors generally which are beyond the sphere of influence of social service agencies.

Similar exercises were initiated in other British cities. A social information study for the city of Manchester, undertaken jointly by the Planning and Social Services Departments, is a good example and illustrates the range of applications that was envisaged for such studies:

- revision of the city's Education Priority Areas;
- supporting claims for Urban Aid applications;
- locating community centres and other social facilities;
- allocation of social workers;
- development of a neighbourhood approach to disseminating welfare rights information; and
- assessing the need for new services or facilities (Manchester City Council 1973, p. 4).

The impending reorganization of local government and the introduction of structure planning in the early 1970s prompted several local authorities at the county level to undertake multivariate census analysis. In the most ambitious of these studies, the methodological approach followed that of the Third Survey of London Life and Labour (Norman 1969) described earlier. It involved a sequence of correlation, principal component and cluster analyses, with the ward as the basic spatial unit. In a large county, the size of data set might well be close to the limit that could be accommodated in a cluster analysis program. For the new Metropolitan County of Greater Manchester, for example, the data matrix amounted to 40 census variables × 442 wards (Joint Working on Structure Plans in the Greater Manchester Area 1972). Here the initial analysis was carried out on 1966 census data, as a prelude to more comprehensive analysis of the 1971 census (Greater Manchester Council 1975).

The 1971 Census Greater Manchester Study is interesting in the way that the clustering process was presented. Instead of concentrating on a single level of clustering, the report presents two levels: an upper level of seven clusters and a lower level of twenty clusters. Two-tier (and sometimes three-tier) classifications became a regular feature of subsequent small area typologies.

5.3 Proprietary geodemographics classification systems

5.3.1 The emergence of ACORN

Up until this point all classification systems had been restricted to one geographical setting: a city or a county. The statistical analysis embodied in a classification was, therefore, relative to local conditions and this made it difficult to make comparisons between the results for different cities or counties. Working at the Centre for Environmental Studies (CES) in the 1970s, Webber began to develop national classifications in which individual clusters could be compared with the national mean for particular census variables. Webber's national classifications were carried out for wards, parishes and local authority areas and were commissioned by the Office of Population Censuses and Surveys (OPCS) (Webber 1977; Webber and Craig 1978).

The ward and parish typology was later renamed ACORN (A Classification of Residential Neighbourhoods) following its first use in the private sector to examine variation in consumer behaviour between area types (see Webber

1989). This resulted from experimental work carried out by McDonald, Baker and Bermingham of the British Market Research Bureau (BMRB). They had cross-tabulated, against ward cluster codes, information collected about product purchases and media use that was assembled for the Target Group Index (TGI: see also Cresswell, Chapter 9, this volume). Significantly better discrimination was achieved, for example in newspaper readership, than had been obtained previously using conventional indicators, such as age or social class. Their paper, at the 1979 Market Research Society Conference, was the marketing community's first exposure to this promising new geodemographic approach.

With the closure of CES in 1979, Webber moved to CACI, then the only commercial organization with an OPCS licence to supply census statistics. At CACI he developed a new enumeration district level 36-cluster classification of neighbourhood types which was marketed as the ACORN system, and included a higher level aggregation to eleven ACORN groups. This original 1971-based ACORN system came to be widely used in both public and private-sector applications (Webber 1985). At CACI, the ACORN classification was linked to the SITE retail catchment characteristic description system. This was enhanced considerably by the availability of cross-tabulations of product consumption by ACORN cluster – a link which provided many marketing personnel with their first practical experience of the application of geodemographics.

A number of methodological criticisms were levelled at the ACORN system by Openshaw *et al.* (1980), primarily relating to inadequacies of the cluster analysis and enumeration-district sampling methods that were employed in its derivation. Openshaw (1983) went on to advocate the use of state-of-the-art methods of cluster analysis in deriving such typologies but underlined the fact that the whole process is extremely subjective, in terms of choice of data, clustering method, number of clusters, and so forth. Openshaw (Chapter 7, this volume) continues to argue that these and other technologies remain subjective, and that there is no single criterion of typology performance – other than its usefulness in practice. Others sounded a note of caution relating to the interpretation of the social processes underlying the patterns that emerged from local policy analysis applications of the system (see, for example, Knox 1978).

A significant breakthrough, which immediately opened up new fields of application, was the linkage of the census-based geography of the ACORN system to the geography of postcodes – and thus to address-based data. The key to this linkage was the grid reference attached to the unit postcode (typically relating to 15 households) that was held on the Postzon file or Central Postcode Directory (CPD: see Martin and Longley, Chapter 2, this volume). This was a file that was originally assembled as part of the Regional Highway Traffic Model (RHTM) exercise, which was concerned with modelling vehicular flows. The link was based on a 'proximal matching' to the grid reference associated with each census enumeration district – a procedure that was error prone. Nevertheless, it provided, for the first time, a means of relating address-based information to the census-based neighbour-hood classification scheme. Use could therefore be made of ACORN for both customer profiling and the targeting of mail to customers or prospects in selected types of neighbourhood. Another important development was the agreement between CACI and CCN that led ACORN codes to be attached to CCN's computer-based version of the Electoral Roll. This made it possible to target selectively the forty or so million names on the Electoral Roll, either by reference to the geodemographic discriminator alone, or cross-tabulated by TGI product/media information. From 1982, this combination was marketed by the Direct Mail Sales Bureau (DMSB) as the Consumer Location System.

5.3.2 The burgeoning geodemographics industry

For a time, CACI dominated the marketplace. However, partly as a result of the movement of former CACI staff, a number of competing products emerged following the release of 1981 census data. Pinpoint, set up in 1983, was one such competitor. Its projects went beyond the direct use of neighbourhood classifications by developing customized products. An example was KIDS – a system for targeting households with high concentrations of children, using raw census statistics (Sleight 1993a). Pinpoint subsequently went on to produce its PiN ED classification system, which first appeared in 1985. During this period there was further fragmentation, realignment of personnel and growth in the geodemographics market. A group of staff left CACI in 1985 to establish Sales Performance Analysis (SPA), based in Leamington Spa, and specialized in the use of geodemographic analysis methods mounted on a PC (see Cresswell, Chapter 9, this volume). In the same year Webber joined CCN, part of the Great Universal Stores (GUS) group, where he developed the system known as MOSAIC, which was released in late 1986. A novel feature of this typology was that it was not based solely on census data. It drew upon data-sets to which CCN had access as a result of its credit referencing activities, including credit applications and County Court Judgements (CCJs) by postcode. Whereas most geodemographic systems were developed by in-house teams of marketing analysts, there was one notable exception. Super Profiles, launched in early 1986, was the culmination of a joint project undertaken at the Universities of Newcastle and Liverpool. It was able to build upon a considerable amount of research on clustering methods and experience in constructing area classification systems in a planning context. FiNPiN was launched at about the same time as MOSAIC – as the first 'market-specific' classification. In this case, Pinpoint used data from National Opinion Poll's (NOP) Financial Research Survey (FRS) to identify census variables which were strongly correlated with financial activity. They then produced a classification, using these census indicators, in which similar financial data were used to describe cluster characteristics in terms which would be familiar to prospective financial-sector users.

With several products in the marketplace, there was a great deal of competition between system vendors, much of which was based upon the claims made concerning the ways in which value could be added to clients' data by using it in combination with both the geodemographic discriminator and other data.

5.3.3 An overview of the principal 1981 census-based geodemographic systems

Four main proprietary geodemographic systems emerged in the mid-1980s, prompted by the availability of 1981 census data. The distinguishing characteristics of the four systems – ACORN (CACI), PiN (Pinpoint), MOSAIC (CCN) and Super Profiles (CDMS) – will now be examined.

Each of the systems took as its starting point enumeration district data from the 1981 census. Each was hierarchical in so far as a finer, basic level of area-type description was summarized at a higher level (or higher levels) of aggregation – primarily to simplify the description of area types which could be used to provide a summary breakdown of variation in behaviour of interest. The ways in which these geodemographic systems have been used in UK market analysis are described in Cresswell (Chapter 9, this volume).

Differences are apparent in the choice of variables, in the clustering methods used and in the number of clusters produced. Further differences are found in the use made of data from sources other than the 1981 census in the derivation and description of cluster characteristics, in the options offered for the presentation and manipulation of related data and the computer systems upon which the systems can be used. The principal similarities and differences between the systems are summarized in Table 5.1, which also lists some of the associated products and services that were offered by the system suppliers.

ACORN

As the first system to be developed and marketed in the UK, the post-1981 version of ACORN shared many features with the original 1971-based system. Partly in order to maintain the loyalty of customers familiar with the latter, CACI adapted the 1971 classification rather than devising an entirely new system from scratch. A similar set of 40 census variables was employed in assigning 1981 census enumeration districts to clusters but the number of clusters was increased from 36 to 38 to reflect the emergence of new area types. The 11 ACORN Groups provide a higher-level summary description of area types.

PiN

Pinpoint emerged as the first rival to CACI with the launch in 1985 of its PiN (Pinpoint Identified Neighbourhood) system based upon the analysis of a larger number of census variables (104) and the identification of 60 area types, which were grouped to 25 and 12 cluster levels of aggregation. PiN was

Table 5.1 Comparison of geodemographic system characteristics

	ACORN	Super Profiles	PiN	MOSIAC
Source/supplier	CACI	CDMS	Pinpoint[a]	CCN
Number of area	–	150	60	58
types	39	37	25	–
	12	11	12	12
Variables				
Census	40	55–65	104	38
Other	–	19 (TGI)		16 (Credit)
Areal units	ED	ED	ED	Postcode
Features	–	Affluence ranking	Wealth indicator	Postal geography
Associated products	InSite/Branch Modeller MONICA WORKFORCE SPECTRA	Statsfile Rollcount Insight Database modelling DBM1	FiNPiN GEOPiN PAC PRN CENTRY LUPiN	Prospect profiling scoring selection TACTICS MACROMAP

[a] See Birkin's discussion of current organizational developments in Section 6.2.1.

marketed within the GEOPiN micro-mounted package based upon the PC-ARC/INFO geographic information system (see Maguire, Chapter 8, this volume).

MOSAIC

In 1986 two further systems were launched. MOSAIC was produced by CCN Systems. It was distinctive in that it was based upon a classification of postcode-based areal units instead of enumeration districts to form 58 neighbourhood types. In deriving the typology, CCN used a total of 54 variables. These included 38 from the census, together with others drawn from the Electoral Roll, Post Office Address File (PAF) and credit rating measures obtained from County Court Judgement records (see Webber 1989).

Super Profiles

The final system described here is Super Profiles marketed by CDMS (part of the Littlewoods Organization), which acquired the original launch company, Demographic Profiles Limited, in 1987. Development work drew upon the results and experience gained in the course of an Economic and Social Research Council (ESRC)-funded project, carried out by Openshaw et al. (1985) at the University of Newcastle. The project was concerned with the use of state-of-the-art methods (described in Charlton et al. 1985) in the derivation of a series of national classifications based on 1981 census Small Area Statistics

(SAS). The Super Profile classification was selected from many generated using 55 SAS variables, based on a systematic evaluation of typologies ranging from 5 to 1000 clusters using 'live trading data' drawn from a national database. It was concluded that the best level of consumer behaviour discrimination was achieved using 150 or fewer clusters. After adding a further 10 SAS variables (see Brown 1988), the 150 Super Profile clusters were labelled according to an 'affluence ranking', and were subsequently grouped, primarily to facilitate description, to form 37 'Target Markets' and 11 'Lifestyles' (see Brown and Batey 1987a; 1987b; 1987c).

A clear distinction can be drawn between the above forms of general-purpose classification scheme and other more market-specific forms of typology. The development of Pinpoint's FiNPiN (or Financial PiN) serves to illustrate the principles used in pursuing this form of more specialized classification scheme. It was derived from links established between census data and details of the use and ownership of financial service products by 30 000 households in NOP's Financial Research Services survey. A total of 58 census variables were observed to correlate strongly with these activities. These variables were then used in constructing a new enumeration district classification consisting of 40 area types, which were grouped first into 10 neighbourhood types. These in turn were grouped into 4 main categories – 'financially active' (20 per cent of GB households); 'financially informed' (26 per cent of GB households); 'financially conscious' (27 per cent of GB households); and 'financially passive' (27 per cent of GB households). In comparison with the general-purpose geodemographic discriminators, the neighbourhood types developed in FinPin have a much more explicit market orientation (Beaumont 1991), in that the labels of the latter reflect the entire range of financial activity involvement.

5.4 1991 census-based geodemographic systems

5.4.1 Design considerations: the example of Super Profiles

By the early 1990s, all the major providers of geodemographic systems had begun development work on 1991 census-based products. Whereas the original systems had to a large extent been technology-led, the new classifications were to be influenced much more by the requirements of users. The geodemographics industry had matured substantially and those who were involved in marketing and applying the systems had a much clearer view about what they needed (Batey and Brown 1994; Brown and Batey 1990; 1994).

When CDMS reviewed the existing Super Profiles in 1991, a firm decision was taken to retain some of the distinctive features of the system: the three-tier classification based on Clusters, Target Markets and Lifestyles; the large number of clusters (150) at the most detailed level; and an affluence ranking of these 150 Clusters. The three-tier structure, it was felt, enabled customers to select the level appropriate for their particular application, scale of activity and degree of sophistication in targeting. The upper limit on the number of clusters (150) should not be increased significantly, because of the risk that the product would prove both intimidating and unwieldy. The affluence ranking would also be retained, even though it was recognized that there were limitations in relying solely on census variables as a proxy for disposable income. CDMS was aware, for example, that the affluence ranking could be misleading in central London, where patterns of housing tenure, mode of travel to work and car ownership levels were quite distinct from those elsewhere in the country.

CDMS was also aware of some of the geodemographic system's short-comings from a marketing viewpoint. The verbal descriptions used to characterize Super Profiles were concerned entirely with the social, economic and demographic features of residential areas as measured by the census, and gave no impression of the variation in consumer behaviour and lifestyle that was associated with these features. The result was regarded as a rather dry, academic presentation of the classification, instead of a system that could be seen as immediately relevant to a commercial client interested in targeting prospective customers for a particular product or service. Table 5.2 gives three examples of the descriptions of Lifestyles used in the 1986 Super Profiles and compares them with their approximate 1994 equivalents. The labelling of the 1986 groups betrays the urban research origins of census classifications.

A related criticism was that the 1986 Super Profiles made no explicit use of non-census data in its construction. While a small number of TGI variables had been used in the formation of Target Markets and Lifestyles, the benefits of such data had not been fully exploited. Nevertheless CDMS felt that it was important to strike a balance between having a general-purpose classification, that is not tied to a limited range of applications, and having a system that made optimum use of the data that were available for particular targeting purposes. They were attracted to the idea of a general-purpose core classification capable of forming the foundation of a number of variants on the basic Super Profiles system.

CDMS went on to identify a series of design principles that could be used to guide the construction of a new Super Profiles system. These principles included the following:

1. Super Profiles would remain a general-purpose core classification based on census data. A deliberate decision was made to use census data alone in constructing the core classification. This reflected the advantages of the census in terms of comprehensive coverage of the whole of Great Britain, its consistency of data specification and the fact that completion of census returns is obligatory.
2. There would be 'visible' and 'invisible' levels of classification. A distinction was made between an 'invisible' classification intended to provide the basis for further developments, and a series of 'visible' classifications available to the user. These visible classifications would together form the Super Profiles product. The invisible classification, on the other hand, would be a

Table 5.2 Lifestyle pen pictures: selected examples from 1986 and 1994

1986	1994
Lifestyle A: Affluent Minority – This most affluent of the Lifestyles is characterized by large detached, owner-occupied housing which accommodates highly qualified, multi-car-owning professional worker households with few children, in low density, suburban and semi-rural areas from which the majority of workers commute by car and train to office jobs.	Lifestyle A: Affluent Achievers – High income families with a lifestyle to match. Detached houses predominate, reflecting the professional status of their owners. Typically living in the stockbroker belt of the major cities, the Affluent Achiever is likely to own two or more cars, which are top of the range, recent purchases and are needed to pursue an active social and family life. Affluent Achievers have sophisticated tastes and aspirations. They eat out regularly, go to the theatre and opera and take an active interest in sports (such as cricket, rugby union and golf). They are able to afford several expensive holidays each year. Financially aware, with a high disposable income, this group invests in both quoted and privatized companies. They are happy to use credit and charge cards and are likely to have private health insurance. Investments are followed closely in the Telegraph. For more leisurely reading, Hello, Harpers and Queen and Vogue are likely to be found in the home of the Affluent Achiever.
Lifestyle B: Metro Singles – Typified by young single, well qualified, professional and white-collar workers, with some single elderly, living in small, furnished and unfurnished rented flats, often lacking in basic amenities, in areas of ethnically mixed population with a high residential turnover, well-served by rail and tube, resulting in an unusually low level of car ownership and use for work travel.	Lifestyle E: Urban Venturers – This cosmopolitan, multiracial group resides in areas of major cities which are undergoing gentrification but still retain a significant proportion of poorer quality housing. These young adults live in terraced houses or flats and have high levels of disposable income, which is spent on eating out, expensive holidays, keeping fit, going to pubs, clubs, concerts and the cinema. Close to where the action is, there is little need for a car; the bus, tube and train are preferred means of transport. They read about their interests in magazines such as Time Out and Cosmopolitan and keep up to date with current affairs in the more liberal broadsheets – the Guardian and the Independent.
Lifestyle J: Underprivileged Britain – This Lifestyle accounts for those areas in which the worst conditions of social stress and deprivation are concentrated. It is characterized by very large families, including young children, living in cramped and overcrowded conditions in council flats that are in generally good condition, but with the highest levels of unemployment amongst a very poorly qualified, unskilled labour force, with very low car ownership, those in employment thus reliant upon the bus for the work journey.	Lifestyle J: 'Have Nots' – Single parent families, living in cramped, overcrowded flats is the everyday reality for this group which is composed of young adults with large numbers of young children. These are the underprivileged who move frequently in search of a break. However, with two and a half times the national rate of unemployment, and with low qualifications, there seems little hope for the future. Most are on Income Support, and those who can find work are in low paid, unskilled jobs. There are very few cars and little chance of getting away on holidays. Recreation comes mainly from the television and the take up of satellite and cable TV is high. Betting is also popular, particularly on greyhound racing. The Sun and the Mirror are the most popular newspapers.

highly detailed classification – probably consisting of several hundred clusters – capable of being augmented and updated with other data, not from the census.

3. A 'cocooning' process would be used to enable layers of new data to be added. The process of constructing a classification system starting from the core, general-purpose classification resembles that of a cocoon. It involves the building of layers of new data around the core, each layer providing additional information to be used in extending the clustering process and in enhancing the descriptions of the clusters that emerge. This 'cocooning process' is illustrated in Figure 5.2, which shows how census descriptors are

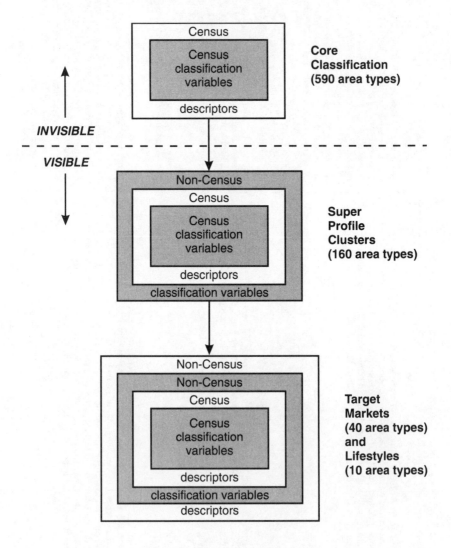

Fig. 5.2 The cocooning process

'wrapped around' the original census classification variables, to be followed by a layer of non-census classification variables (e.g. from the Electoral Roll or from Littlewoods trading data) and, finally, by a layer of non-census descriptors (most notably, TGI data). There would also be opportunities to discard these outer layers of data to allow new data to be added so that the classification could be readily updated.

4. User-friendly pen pictures would be devised. In describing the distinctive features of particular clusters, a serious attempt would be made to couch descriptions in terms that were more meaningful to potential users (see Table 5.2). At the same time, special care would be taken to ensure that the labels adopted for clusters were not misleading, referring only to minority elements within residential area types.

5. Other important strengths of the original Super Profiles system would be preserved. That is to say, there would be three levels of classification; the classification would be detailed and would entail approximately 150 clusters; and there would be an affluence ranking of clusters.

Figure 5.3 summarizes the main aspects of the overall approach that have been pursued in the development of the new system: in particular, in seeking to meet the above requirements. An important distinction is made between the so-called 'invisible' classification and the 'visible' classifications made available to the user. The basic principle centres around the development of the 'invisible' classification, which consists of a relatively large number of clusters that are derived using census-based indicators. This classification is not seen by the typology user. It consists of more clusters than most users would find useful but can be aggregated to produce the 'higher' levels of area type.

An innovative feature of this scheme is that indicators derived from the Electoral Roll and TGI for each of these invisible clusters can be used to enhance the description of the higher-level clusters. As illustrated in Table 5.2, the enhancements can be expressed in terms (shown in italics in Table 5.2) which are more familiar to the marketing specialists than were the census-based descriptions of the 1981 system. The indicators used for this purpose can reflect variation in consumer behaviour and media preferences as well as Electoral Roll information relating, for example, to household structure.

The new system meets a basic criticism that has frequently been made of the 1981 census-based classifications. The credibility of these systems declined as the years passed, and little consideration was given to the effects of what were likely to have been important changes in both the distribution of people and their behaviour. In the new system non-census information is used, from the Electoral Roll and from the Target Group Index (TGI), both to enhance the description of cluster characteristics and to provide a basis for the future updating of classifications to reflect changes in these characteristics. The new Super Profiles has therefore been able to take advantage of experience that has been gained in using the original 1986 version of the classification (see Evans and Webber 1994 for a similar account with respect to the re-development of the MOSAIC system).

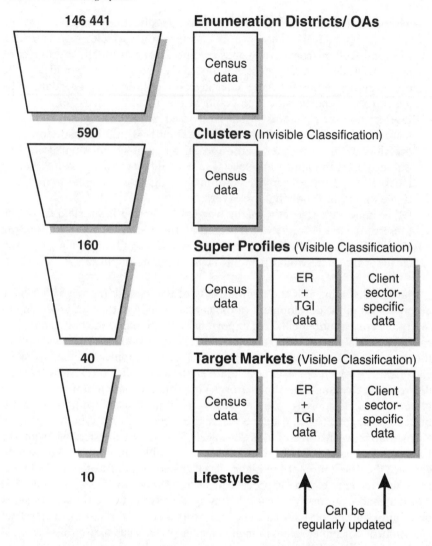

Fig. 5.3   The new Super Profiles: schematic diagram

New features of the 1991 census have made possible further improvements in all of the major geodemographic systems. For example, among the more obvious enhancements has been the use made of information that was recorded for the first time in the 1991 census, such as that relating to dwelling type and the 'life stage' represented by household structure. The developers of new systems have also shared the benefit of another fundamental improvement over the 1981 census. This has been achieved with the release of the enumeration district-postcode directory, which provides a much more accurate link between census and postal geography than the crude 'proximal matching' relationship employed in the past (see Gatrell 1989, and Raper *et al.* 1992).

5.4.2 Highlighting and exploiting variations between clusters

Users of geodemographic systems are able to derive a number of benefits from the multi-dimensional nature of an area typology. Some of these benefits relate to the rich description of the distinguishing features of the types of area identified in the classification that can be assembled. Other benefits can flow from recognizing how the properties of individual clusters might account for variation in consumer behaviour or patterns of events between clusters. Even without a direct behavioural explanation, awareness of the degree of discrimination or difference between clusters can itself be exploited in making a decision about a course of action.

Understanding the complex inter-relationships between cluster attributes can be aided in a number of ways. Among the simplest is an index table which provides a convenient means of presenting information with which to compare the characteristics of clusters. Such a table relating to the ten new Super Profile Lifestyles is presented as Table 5.3. This Table provides a basis for comparing the ten Super Profile Lifestyles with respect to 120 1991 census variables. Column 1 gives a description of the census variable and column 2 shows the percentage of the relevant population base category that is in this group (e.g. 6.64 per cent of all individuals are in the category '% persons aged 0–4', 10.61 per cent of all households are in the category '% residents in households age 16–24, no children 0–15', etc.). Column 3 denotes whether each census variable was used in the derivation of the Super Profile classification. The remaining columns set out index values which compare the Lifestyle cluster mean values for an individual variable with the corresponding mean for the country as a whole, a value greater than 100 indicating a cluster mean greater than the national mean. The Super Profile Lifestyles are set out in descending order of affluence, moving from Lifestyle A through to J. This form of analysis can be extended to include a wide range of non-census variables.

The information set out in this form of index table was used to produce the simple 'pen picture' verbal descriptions of the distinguishing features of each Lifestyle cluster of the type featured in Table 5.2. This indicates that the cluster descriptions draw upon a wider range of information that appears in an extended version of Table 5.2, which, in addition to variables derived from the census, includes information obtained from the Electoral Roll, Littlewoods trading data and, in particular, from the Target Group Index (TGI). The latter, to which reference has been made above, is the outcome of a regular survey of the product consumption and media preferences of approximately 24 000 respondents.

Three of the new pen pictures (Lifestyle A: Affluent Achievers; Lifestyle E: Urban Venturers, and Lifestyle J: 'Have Nots') are illustrated in Table 5.2, in which they are set alongside the nearest equivalent pen pictures from the 1986

Table 5.3 Super Profile Lifestyle index table

Var no.	Variable name	Global mean	CI[a] va	Super Profile Lifestyle[b]									
				A	B	C	D	E	F	G	H	I	J
1	% Persons Aged 0–4	6.64	*	76	69	91	122	114	85	76	83	117	148
2	% Persons Aged 5–14	12.32	*	105	85	102	116	92	100	62	86	119	119
3	% Persons Aged 15–24	13.84	*	97	86	96	99	122	92	93	90	108	112
4	% Persons Aged 25–44	29.19	*	91	85	103	121	122	95	89	85	97	96
5	% Persons Aged 45–64	21.97	*	122	121	108	83	75	119	93	111	91	83
6	% Persons Aged 65–74	9.02	*	90	129	88	66	67	96	145	142	93	88
7	% Persons Aged 75+	7.01	*	90	137	84	59	79	91	205	131	75	78
8	% Residents 18+	76.28		99	104	100	95	100	100	107	104	95	93
9	% Residents in Households Age 16–24 No Children 0–15	10.61		100	88	100	102	134	90	100	86	98	96
10	% Residents in Households Age 16–24 Children 0–1	5.32	*	85	68	84	102	108	86	67	89	143	171
11	% Residents in Households Age 25–34 No Children 0–15	9.16	*	71	74	88	116	199	80	118	73	79	96
12	% Residents in Households Age 25–34 Children 0–4	7.14	*	68	65	102	146	111	80	73	81	117	125
13	% Residents in Households Age 25–34 Youngest Children 5–10	2.49	*	49	56	90	135	83	68	63	100	168	163
14	% Residents in Households Age 25–34 Youngest Children 11–15	0.41		46	53	75	117	90	74	64	109	183	180
15	% Residents in Households Age 35–54 No Children 0–15	18.01		113	105	109	100	87	115	88	97	94	90
16	% Residents in Households Age 35–54 Children 0–4	3.45	*	133	94	106	118	119	122	63	70	87	90
17	% Residents in Households Age 35–54 Youngest Children 5–10	5.57	*	141	101	119	126	77	118	58	76	93	77
18	% Residents in Households Age 35–54 Youngest Children 11–15	5.34	*	138	102	123	121	61	116	59	85	101	77
19	% Residents in Households Age 55 – Pens Age Working/Retired	6.69	*	132	132	112	79	72	127	89	107	82	71
20	% Residents in Households Age 55 – Pens Age Unemp/EconInact	3.00		86	109	92	79	56	102	95	137	123	116
21	% Residents in Households Pensionable Age – 74	14.70		92	124	90	70	67	97	129	136	98	95
22	% Residents in Households Age 75+	8.11	*	85	128	85	62	80	87	188	128	81	86
23	Average Household Size	2.45		109	97	106	108	93	105	80	95	106	98
24	Average Household Size	2.21	*	92	65	73	87	185	112	48	76	141	147
25	% Households Married Couples	78.98		112	109	107	102	89	109	96	100	91	74
26	% Households 2 Aged 16+ and 1+ Dep Children	36.53	*	103	80	108	125	92	98	67	82	114	113
27	% Females of Child-bearing age 16–44	41.22		95	84	101	115	123	95	86	84	101	104
28	Infant : Young Woman Ratio	13.05		78	69	92	123	115	89	74	81	116	144
29	% Lone Parent Families	4.18	*	41	43	57	89	102	43	67	87	167	285
30	% Lone Female Pensioner Households	11.88	*	74	113	82	64	78	67	178	136	91	98

#													
31	% Resident Population African Born	0.60		78	48	68	51	433	37	55	26	46	158
32	% Resident Population Caribbean Born	0.48		29	25	45	42	451	13	46	30	68	247
33	% Resident Population Indo/Pakistani Born	0.94	*	67	45	61	47	526	23	62	33	49	84
34	% Resident Population Non-Commonwealth/EEC Born	2.31	*	138	87	72	73	282	73	93	49	49	93
35	% Residents in Households – Black	1.61	*	19	17	35	38	419	7	42	29	74	316
36	% Residents in Households – Indians/Pakistanis/Bangladeshis	2.72	*	45	25	51	42	619	4	47	26	53	79
37	% Residents in Households – Chinese + other	1.16		90	50	73	70	359	20	75	36	60	148
38	% Residents in Households – New Commonwealth	3.07		65	44	61	52	475	27	61	32	53	126
39	Ratio Non-White Population to Population Non-UK Born	81.21		49	43	73	75	139	17	63	64	106	139
40	% One Year Migrants	9.61		85	84	73	103	167	97	136	77	83	111
41	% Wholly Moving Households Moving Between Districts	33.36		137	128	110	108	114	152	93	72	61	54
42	% Wholly Moving Households With Dependent Children	36.84	*	120	86	111	118	68	106	54	98	138	130
43	% Moves by Wholly Moving Households by Persons Aged 60+	6.83	*	65	151	75	52	47	69	232	191	74	74
44	% OAP Migrants as % OAPs	4.05		96	107	75	92	104	108	160	94	67	88
45	% Migrants who are Non-White	5.42		47	28	51	47	506	8	51	29	61	164
46	% Single Worker Households with No Children	10.57	*	68	77	73	89	201	86	142	71	73	124
48	% Households with 2+ Adults Dual Income No Children (DINKYs)	20.91		116	105	120	117	105	119	83	88	85	59
49	% Married Females Working Housewives	53.16	*	104	95	111	115	100	96	90	89	93	81
50	% Economically Active Self Employed	11.52		148	143	97	81	90	247	102	81	57	48
51	% Economically Active Part-Time Workers	16.20	*	111	111	112	103	62	93	90	109	101	86
53	% Employed Residents 16+ : Cler, Sec + Sales working 41+ hrs/week	1.62		104	100	103	99	127	89	109	82	80	94
54	% Workforce Male	57.47	*	100	100	98	98	99	106	99	100	101	101
55	% Residents Aged 16+ Students	3.83	*	167	111	101	87	147	120	70	64	69	77
56	% Residents Aged 18–24 Students : Term-Time Address in ED	11.79		162	119	100	79	159	90	84	63	55	77
57	% Residents 16–24 on Government Scheme	3.01		57	75	90	96	63	104	80	134	145	146
58	% Economically Active Males Unemployed	11.17	*	44	57	59	69	129	50	111	113	154	249
59	% Qualified Residents 18+ and < 60/65: Unemployed	3.37		75	83	78	75	149	90	119	105	137	243
60	% Households in Furnished Rented Property	21.84		88	90	99	103	97	78	91	107	114	95
61	% Unemployed Males (or Government Scheme) Previous Job : Construction	20.97		70	86	99	104	78	80	102	110	110	109
62	% Unemployed Males (or Government Scheme) Previous Job : Service Industry	45.33		119	109	102	97	114	103	107	88	83	99
63	% Unemployed Males (or Government Scheme) Previous Job : Professional or Managerial	16.41		234	169	135	107	133	157	102	64	46	47

Table 5.3 contd.

Var no.	Variable name	Global Mean	CI[a] va	Super Profile Lifestyle[b]									
				A	B	C	D	E	F	G	H	I	J
				70	86	106	106	82	88	95	112	113	101
64	% Unemployed Males (or Government Scheme) Previous Job : Craft or Related	29.69	*										
65	% Households with 1–3 Rooms	14.81	*	31	54	41	52	205	40	209	102	70	184
66	% Households with 4–6 Rooms	70.64		78	97	111	109	84	85	84	109	119	98
67	% Households with 7+ Rooms	14.55	*	273	157	102	100	69	232	61	50	36	23
68	Persons per Room	48.91		88	88	99	104	108	88	93	100	111	116
69	% Households Severely Overcrowded > 1.5 persons per room	0.50	*	31	37	32	40	394	71	136	43	71	177
70	% Households Share/Lack Bath	0.92		49	70	40	40	332	113	252	56	56	65
71	% Households Share/Lack Inside WC	0.97	*	51	67	41	49	323	119	222	52	85	71
72	% Households with No Central Heating	20.47	*	40	61	65	80	131	131	123	114	146	148
73	% Households in Property with Central Heating in All Rooms	67.69		122	110	111	105	95	86	93	93	81	83
74	% Households in Owner Occupied Property	66.39		132	124	131	120	92	104	89	84	75	34
75	% Households in Owner Occupied Property – Bought Outright	23.93	*	138	166	131	81	74	140	108	94	65	25
76	% Households in Owner Occupied Property – Buying	42.46	*	129	101	131	142	102	84	78	79	80	40
77	% Households in Council/New Town Rented Property	21.44	*	15	32	31	55	49	25	96	166	196	296
78	% Households in Housing Association Property	3.13		24	48	27	62	193	21	178	100	69	244
79	% Households in With Job Property	1.90	*	109	111	52	79	148	567	89	66	55	72
80	% Households in Unfurnished Rented Property	3.63	*	79	100	69	72	186	294	166	61	89	63
81	% Households in Furnished Rented Property	3.51	*	75	80	54	59	351	86	195	39	44	56
82	% Household Spaces as second and holiday Homes	0.72	*	66	272	32	25	129	650	139	58	11	13
83	% Dwellings Non-Permanent	2.26		92	85	117	122	20	222	91	143	94	64
84	% Unshared Dwellings Detached	20.30	*	279	203	98	108	18	314	50	55	13	5
85	% Unshared Dwellings Semi-Detached	30.02	*	89	115	177	113	41	76	52	127	96	47
86	% Unshared Dwellings Terraced	29.38	*	32	45	65	121	134	32	106	114	194	116
87	% Unshared Dwellings Purpose-Built Flats	16.60	*	25	45	38	43	166	10	211	93	61	287
88	% Unshared Dwellings Bedsits	0.02	*	68	71	34	41	383	70	236	44	46	61
89	% Total Dwelling Stock Vacant	4.64	*	76	89	67	84	179	131	144	69	76	124
90	% Households without a Car	33.36	*	34	58	59	67	124	34	139	125	138	192
91	% Households with One Car	43.51		94	109	113	114	100	101	93	97	97	68

No	Variable	*	Mean	5	10	7	19	24	8	21	18	9	39
92	% Households with Two Cars	*	19.15	197	138	133	125	65	179	57	67	51	26
93	% Households with 3+ Cars	*	3.99	242	155	131	100	48	254	50	64	42	20
94	% Non Single Person Households with No Car		19.26	23	42	50	66	144	24	136	133	169	253
95	% Skilled Man and Unskilled Worker Households With No Car		24.72	32	42	40	61	145	37	133	98	128	204
96	% Skilled Man and Unskilled Worker Households Owner Occupiers		67.20	121	118	131	122	98	84	95	95	86	43
97	% Persons Work Travel Mode : Travel by Car	*	60.98	117	112	110	112	72	100	87	97	87	66
98	% Employed Residents 16+: Commute Entirely by Car (all workers)		28.59	120	115	104	112	75	103	92	93	82	68
99	% Persons Work Travel Mode : Travel by Bus	*	9.88	36	49	78	91	125	17	91	123	173	240
100	% Persons Work Travel Mode : Travel on Foot	*	11.60	48	70	76	84	113	57	168	130	143	153
101	% Persons Work Travel Mode : Working at Home		4.92	149	151	73	58	86	454	120	67	41	47
102	% Persons Work Travel Mode : Travel by Train	*	5.78	112	65	104	55	326	24	76	37	59	134
103	% Agricultural Workers	*	1.96	109	208	40	37	15	1082	44	61	33	25
104	% Energy/Water Workers		4.72	88	87	108	120	59	62	72	126	135	83
105	% Manufacturing Workers		17.88	74	80	96	107	80	59	90	129	153	106
106	% Service/Distribution Workers	*	67.16	109	103	101	98	112	83	106	89	83	100
107	% Armed Forces SEG 16		0.62	66	77	64	176	226	88	78	55	47	46
108	% Employers and Managers SEG 1+2	*	14.38	174	129	108	102	104	111	98	66	46	43
109	% Professional SEG 3+4		4.46	220	142	95	88	144	103	87	45	28	31
110	% Other White Collar SEG 5.1 5.2 6 7		36.97	111	108	114	107	108	75	103	85	74	77
111	% All White Collar	*	55.82	136	116	111	104	110	87	100	77	63	65
112	% Skilled Manual SEG 8+9	*	14.80	48	72	100	109	73	58	94	138	156	120
113	% Semi-Skilled Manual SEG 10		10.72	41	65	78	96	86	57	102	143	176	153
114	% Unskilled SEG 11	*	5.55	35	63	68	84	74	71	108	151	172	200
115	% All Manual		31.07	43	68	86	100	77	60	99	142	166	146
116	% Non Prof/Self Employed SEG 12 13 14	*	6.49	97	122	109	89	86	228	112	97	75	66
117	% Qualified Persons Aged 18+		13.43	208	140	101	96	147	129	85	45	28	33
118	% Qualified Male Residents 18+		15.55	218	142	101	96	138	118	84	43	26	31
119	% Qualified Female Residents 18+		11.52	194	136	101	96	157	141	88	47	31	36
120	% Residents 18+ Higher Degrees		0.93	255	144	77	72	214	117	77	24	14	32

Cluster Size =   5   10   7   19   24   8   21   18   9   39

a * denotes variable used in cluster derivation

b global mean set to 100 and Lifestyles in rank sequence of affluence

Super Profiles classification. Highlighted in italics, for the 1994 Lifestyles, are the elements of the pen pictures which draw upon consumption and lifestyle information that is derived from non-census sources. The table shows clearly how much more user-friendly the new pen pictures are in comparison with their 1986 near equivalents.

However, care should always be exercised in interpreting the values that appear in an index table of the type that appears as Table 5.3. To illustrate the point, see Table 5.4, which contains information relating to an extract of variables from Table 5.3 for the two lifestyles at opposite extremes of the affluence spectrum – Lifestyle A: Affluent Achievers and Lifestyle J: 'Have Nots'. Contrasts between the two types of area are evident from the very high and low index values. For example, the mean values of 'Lone Parents' and 'Male Unemployed' for Lifestyle A enumeration districts are relatively low (41 and 44 percent of the national mean) in comparison with the high values for enumeration districts classified as falling in Lifestyle J (almost 3 and 2½ times the national mean, respectively). Similarly, in the case of 'Households with 3+ Cars', Lifestyle A can be seen to have a mean value which is almost two and a half times the national mean. However, in this case, it should be recognized that the national mean itself is only 4.0 per cent. This implies that the mean for Lifestyle A enumeration districts is itself only 12 per cent. This, in turn, implies that although the average representation of households owning more than one car is high in this type of area, this still represent only a minority of households – 88 per cent are, on average, NOT multi-car-owning. Birkin (Chapter 6, this

Table 5.4 Extract from Super Profile Lifestyle index table: comparison of Lifestyle mean values with national mean values

| Variable name | National mean | (National Mean = 100) | |
| | | Lifestyle A 'Affluent Achievers' | Lifestyle J The 'Have Nots' |
	(%)	Index value	Index value
Children 0–5	6.6	76	148
Older adults 45–64	21.8	122	83
Lone parents	4.2	41	285
Male unemployment	11.2	44	249
Property 1–3 rooms	14.8	31	184
Property 7+ rooms	14.6	273	23
Owner occupied	66.4	132	34
Rented from council	21.4	15	296
Detached	20.3	279	5
Purpose-built flats	16.6	25	287
No central heating	20.5	40	148
No car	33.4	34	192
3+ Cars	4.0	242	20
Professional	4.5	220	31
Unskilled worker	5.6	35	200

volume) presents a similar analysis with respect to household propensity to patronize holiday camps.

As well as facilitating cluster description, the presentation of information in the form of an index table can also highlight the extent to which the sales of a product, incidence of a health condition or occurrence of a particular type of crime is under- or over-represented in areas falling in specific clusters. This variation in representation, or in penetration rates within individual types of area, is the key to many practical applications of geodemographics in which recognition of a significant pattern may be the trigger for some type of action. This might take the form of a targeted mailing, the opening of a new retail outlet or the deployment of additional health service or police resources (see Brown 1991 and Brown *et al.* 1991; Beaumont 1991; Beaumont and Inglis 1989; Carroll 1992; Flowerdew and Goldstein 1989; and Birkin, Chapter 6, this volume, for more examples of the ways in which geodemographic systems may be employed). Here we provide two simple examples which serve to illustrate the potential for this type of application – the use of geodemographic indicators in model-based market analysis is explored and illustrated more fully in Cresswell (Chapter 9, this volume) and Clarke and Clarke (Chapter 10, this volume).

In the first example, Table 5.5 records the index values of a number of the TGI variables that were used in producing the pen pictures for the Lifestyles. Some relate to attributes associated with wealth, such as the ownership of stocks and shares, while others record the extent to which readership or purchase of newspaper and special-interest magazines varies between clusters.

Table 5.5 Summary table of variable index values of selected TGI variables by Super Profile Lifestyle

Variable name/description	Global mean	Super Profile Lifestyle									
		A 1	B 2	C 3	D 4	E 5	F 6	G 7	H 8	I 9	J 10
Watch golf on TV	20.6	130	114	109	100	80	94	99	94	82	74
Watch wrestling on TV	13.9	66	75	86	102	99	81	94	113	130	144
Two holidays per year	27.3	146	123	117	99	101	89	90	80	72	60
Stocks and shares	21.8	165	139	118	91	100	131	98	76	49	40
Private health insurance	15.5	197	131	115	95	99	125	92	65	53	40
Daily Telegraph	8.0	229	163	106	72	114	150	99	52	30	37
Daily Mail	11.6	143	134	121	101	118	104	94	79	46	42
Daily Mirror	18.3	49	67	93	109	103	51	84	127	135	129
The Sun	23.0	51	62	88	105	102	82	95	119	133	152
Fishing magazines	2.4	60	64	118	113	70	34	79	112	118	162
Boating magazines	1.1	172	145	76	105	119	273	123	64	22	37
Good Housekeeping	2.9	247	175	89	73	95	149	95	60	30	27
Practical Parenting	2.9	63	94	68	85	62	154	113	138	132	123

It is interesting to note the contrasting distributions of some of the latter between the more affluent (A, B and C) and the less affluent (H, I and J) ends of the Lifestyle spectrum. Reference to these 'lifestyle' variables clearly adds to the depth of the pen-picture descriptions that can be applied to the residents of areas in which an individual Lifestyle is over- (or under-) represented.

In the second example (see Table 5.6), a more explicit indication is given as to how index values may be derived in a slightly different way from the above 'cluster mean values' by comparing, for each cluster, the proportion of a client's file (in this case those giving donations to charities) living in areas assigned to each cluster with the proportion of the population as a whole found in each cluster. It is evident that there is a marked contrast between the high rates of penetration achieved in some Lifestyles (e.g. A and B) in comparison with others (Lifestyles I and J) – a contrast that is likely to be still more marked in an analysis conducted at the finer-grained 40–cluster Target Market level of classification. In these circumstances, the information could be used, for example, to identify the types of area (and thus postcodes and addresses) in which a mailing of material requesting charitable donations might be concentrated or, sometimes more importantly, those areas to be excluded from an otherwise blanket mailing campaign.

5.5 Concluding comments

This chapter has traced the evolution of the field which has come to be known as geodemographics from its origins in the area classification work of urban researchers to the design considerations which have influenced the development of the most recent geodemographic systems. The emphasis in the earlier sections of the chapter was upon underlining the parallels between the use of area classifications to meet the policy-making interests of the urban

Table 5.6 Super Profile Lifestyle penetration rates: charitable donations

		Percentage of client's file cases	Percentage of national population	Index client's file/ national population
A.	Affluent Achievers	27.0	11.1	243
B.	The Thriving Greys	21.2	13.3	160
C.	Settled Suburbans	9.4	12.1	77
D.	Nest Builders	7.0	14.0	50
E.	Urban Venturers	13.0	11.3	116
F.	Country Life	5.3	3.0	177
G.	Senior Citizens	7.8	8.7	90
H.	The Producers	6.0	13.5	44
I.	Hard-Pressed Families	1.0	5.0	19
J.	The Have-Nots	2.0	7.8	26

planner and the contemporary concerns of the marketing analyst in generating up-to-date information that is actionable. The focus then shifted to the development of proprietary geodemographic systems in the 1980s which took advantage of the critical link between the census geography of the classifications and the postal geography, which plays a vital role in many fields of application. Following the release of 1991 census data, the redevelopment of the major geodemographic targeting systems has provided a welcome opportunity to reflect on the advantages and shortcomings of these systems from the perspective of both the user and the marketing executive. The final section has distilled some of this practical experience and outlined how lessons learnt have been used to redesign a geodemographic product, in this case Super Profiles. Finally, brief reference has been made to one of the prerequisites of geodemographic system use – an understanding of the distinctive features of individual clusters – and to the ways in which variations in cluster characteristics and response or penetration rates can be exploited in system applications.

No attempt has been made to review or reference the many ways in which geodemographic systems can be applied and combined with other data, products and procedures. Some aspects of these issues are examined by Birkin (Chapter 6, this volume), while the use of geodemographics as input information to market analysis models is investigated by Cresswell (Chapter 9, this volume) and Clarke and Clarke (Chapter 10, this volume).

Abundant processing power now means that computationally complex technical innovations or typology enhancements, which were simply infeasible or unachievable over a realistic time scale ten years ago, can now be explored and tested relatively easily. Indeed, the processing issues that are raised in this field are now much less important than the need to gain the commitment of managers to ensuring that full advantage is taken of the opportunities that are available to add value to the mountains of data on which they sit. Many GIS-based tools exist which can enable vital clues to new or expanded business to be wrung from information about the behaviour of existing customers and links between these records and aggregate descriptors like Super Profiles. The challenge for the future is to demonstrate how this potential can be realized.

6

Customer targeting, geodemographics and lifestyle approaches

Mark Birkin

6.1 Introduction

It is necessary to begin by saying what we mean by 'geodemographics', and this is by no means a straightforward task. Interestingly, there is no description to be found in the third edition of the *Dictionary of Human Geography* (Johnston *et al.* 1991), which stands on my bookshelf. This in itself is an eloquent comment on the extent of the recent growth of interest in the topic in the mid- to late 1980s; and perhaps also on the continuing failure of mainstream geography to dirty its hands with applied issues!

Turning to an excellent review of the topic by Brown (1991) we find what we might call the 'narrow' definition:

> Geodemographics has come into use as a shorthand label for both the development and the application of area typologies that have proved to be powerful discriminators of consumer behaviour and aids to 'market analysis'.

These area typologies go under names and acronyms like ACORN and MOSAIC in the UK; and Lifestyles, ClusterPlus and VISION in North America.

A broader and more literal view takes geodemographics to mean any kind of small area demographic analysis. Thus Flowerdew and Goldstein (1989) describe the growth of the geodemographics industry in Canada arising from the provision of small area census information to the private sector through a time-sharing computer. Subsequent development was fuelled by the provision of digital maps to go with these data. Within the 'broad' definition, then, the use of area typologies is only one of a variety of approaches which are relevant to geodemographics.

To try and clarify the major issues, in this chapter we will adopt the narrow definition of geodemographics as relating to the types of systems which provide area typologies. We will follow Beaumont's lead in characterizing the broader field of interest as 'Market Analysis' (Beaumont 1989). The point to be made next is that if we accept these definitions, then in the UK there has historically been very little distinction between the two areas. Geodemographics and market analysis are virtually synonymous. However, this has been true to a rather lesser extent in North America, and perhaps even less so in continental Europe, where geodemographic systems have hardly been seen (but see also Section 6.2 below). Furthermore, and more importantly, there are good reasons to suppose that this will not continue to be the case in the indefinite future. In this chapter we will look particularly at the ways in which market analysis is beginning to benefit from the adoption of a spectrum of approaches which go beyond traditional geodemographics (see also Openshaw, Chapter 7, this volume). The main focus will be on the use of 'lifestyle' data originating from consumer surveys as an alternative to government census data, although there will also be some discussion of the use of mathematical modelling in market analysis.

The remainder of this chapter is structured into three sections. Section 6.2 will consider the geodemographic systems in more detail. In Section 6.3, the concept of lifestyle data will be introduced and we will discuss the ways in which the data may be used in market analysis. In Section 6.4 an assessment of the two different approaches – lifestyles versus geodemographics – will be offered, together with views on future developments within the market analysis 'industry'.

6.2 Geodemographics

It is not considered necessary here to present a detailed overview of the historical development of geodemographic systems (see, for example, Brown 1991 and Batey and Brown, Chapter 5, this volume). Three sets of questions which will be addressed in more detail are the following:

1. What geodemographic systems are currently available in the marketplace?
2. How are geodemographics being applied? What are the strengths and weaknesses of these applications?
3. How can the vendors and users of geodemographic systems respond to problems with the existing systems?

We will be drawing examples primarily from the geodemographics industry in the United Kingdom, while recognizing that the industry is somewhat less well-developed here than in the United States, but considerably more advanced than in continental Europe.

6.2.1 The current status of geodemographics

Geodemographic systems may use a variety of types of data to generate area profiles, but far and away the most important is government census data. In this section we will begin by reviewing the current state of development of the UK geodemographics industry, since it is reasonably developed by international standards. We will then consider the development of pan-European geodemographics.

The most recent UK census was conducted in April 1991, although the bulk of the data have only recently been made available in a digital form. While census data have been gathered and processed by the Office of Population Censuses and Surveys (OPCS), a government office, private-sector users will normally access data through an appointed third-party 'census agency'. Each agency pays OPCS a substantial fee for the right to hold and handle the census data (the size of the sum varies according to the precise use of the data, but a figure of £40 000 is not unusual), and passes on the appropriate royalty payments from clients who use the data to OPCS. In return, the agencies are free to reprocess and package the data in whatever way they feel appropriate, and to make such charges for their census products as they feel the market will stand.

According to the Census Interest Group of the Market Research Society (Leventhal *et al.* 1993), 13 UK organizations are currently registered as census agencies. The organizations are listed and grouped into categories in Table 6.1. In the left-hand column of this table three agencies are listed whose main interest lies in reselling the data with value-added in one of three ways: first by offering the data in a more user-friendly form, e.g. within a Windows environment for accessing the data (Chadwyck-Healey, Claymore); secondly by providing a capability for mapping the data (Chadwyck-Healey, Claymore); and thirdly, by linking between the census and postal geography (Capscan, Claymore). In the right-hand column of Table 6.1 we find three agencies whose primary interest lies in adding value to census data through the provision of some kind of analytical capability. Two of these organizations are represented elsewhere in this volume (Chapters 9 and 10); see also Section 6.2.2 below.

The central column of Table 6.1 shows seven organizations whose primary interest in UK census data is in the production of geodemographic classifications. Note that since the original agency agreements were signed, two of these companies – CACI and Pinpoint – have effectively merged under the CACI banner. Currently, CACI plans to continue to offer both ACORN and Pin as separate geodemographic products, subject to client demand. It is also surprising to observe that only one of the seven players (Equifax) is a newcomer since the last round of geodemographic products based on the 1981 census, although InfoLink and EuroDirect have only joined the market in the late 1980s and early 1990s respectively. Interestingly (in view of what follows in Section 6.3 below) the activities of Equifax, the newcomer, have more to do with Lifestyles than with census data.

CACI's ACORN system was the first and is probably the best-known of the classifications. The 1991 version of ACORN allocates each of the 147 000 or so census enumeration districts in Great Britain to one of six neighbourhood 'Categories' (CACI 1993a). The shorthand labels for these six Categories are 'thriving', 'expanding', 'rising', 'settling', 'aspiring' and 'striving'! For more

Table 6.1 Categorization of 1991 UK Census agencies

Data focus	Geodemographic focus	Analytical focus
Capscan Ltd	CACI Ltd	The Data Consultancy
Chadwyck-Healey Ltd	CCN Marketing	GMAP Ltd
Claymore Services Ltd	CDMS Ltd	SPA Marketing Systems
	EuroDirect Database	Ltd.
	Marketing Ltd	
	Equifax Europe (UK) Ltd	
	Infolink Decision Services Ltd	
	Pinpoint Analysis Ltd	

Source: Leventhal *et al* 1993.

detailed analyses, these Categories may be disaggregated further into 17 ACORN 'Groups'; or, for even more detail, into 54 ACORN 'Types' – see Figure 6.1.

With the fusion of CACI and Pinpoint, CCN's MOSAIC system is the major competitor in terms of commercial market share to ACORN (but see also Batey and Brown, Chapter 5, this volume). In contrast to ACORN, MOSAIC features only two subdivisions into 11 groups and 52 types (Webber 1993). Each neighbourhood receives a rather more evocative description within MOSAIC, for example 'Pebble Dash Subtopia' and 'Bohemian Melting Pot'! (see Table 6.2). How representative these labels are of the actual populations of these areas is an open question to which we will return shortly. One is inclined to wonder how many residents of Bexleyheath would recognize themselves as pebble dash subtopians! Nevertheless, each category is also backed by a more detailed demographic description, a photograph of a typical residence or street, and a textual profile, sometimes known as a 'pen portrait' – see Figure 6.2 for example.

A further distinction between ACORN and MOSAIC is that while the former system is based purely on the classification of census data, the latter incorporates information from a number of other sources – county court judgements, credit activity, Electoral Roll, Postcode Address File, retail accessibility, population density, and company directorships (see Webber 1993, pp. 6–7). The systems developed by Pinpoint and by EuroDirect are also based on census data only, while the others all embrace some external data (cf. Sleight 1993b) – see Table 6.3.

The incorporation of non-census data also allows CCN to produce classifications for each of the 1.5 million unit postcodes in the country. Each postcode has an average of 15 households, compared with an average of 180 for census enumeration districts (EDs). Thus MOSAIC is available for a much finer geography than ACORN. The philosophy here is to use the additional, non-census data sources described above to test for similarity between a unit postcode and the ED in which it falls. If the postcode appears to be similar to the ED, or if there are very few data at the postcode level, then the postcode will probably be assigned to the same category as the ED. On the other hand, if

ACORN Categories	ACORN Groups		ACORN Types		Percentage of households in GB	Corresponding social grades	Corresponding 1981 ACORN Types
A **THRIVING**	1	Wealthy Achievers, Suburban Areas	1.1	Wealthy Suburbs, Large Detached Houses	2.2%	AB	J36, B6
			1.2	Villages with Wealthy Commuters	2.8%	AB	J35, A1
			1.3	Mature Affluent Home Owning Areas	2.7%	ABC1	J34, J36
			1.4	Affluent Suburbs, Older Families	3.4%	ABC1	B5, J34
			1.5	Mature, Well-Off Suburbs	2.9%	ABC1	J34, J33
	2	Affluent, Greys, Rural Communities	2.6	Agricultural Villages, Home Based Workers	1.5%	ABC2D	A1, A2
			2.7	Holiday Retreats, Older People, Home Based Workers	0.7%	ABC2D	K37, A1
	3	Prosperous Pensioners, Retirement Areas	3.8	Home Owning Areas, Well-Off Older Residents	1.5%	ABC1	K37, J33
			3.9	Private Flats, Elderly People	1.3%	ABC1	K38, K37
B **EXPANDING**	4	Affluent Executives, Family Areas	4.10	Affluent Working Families with Mortgages	1.8%	ABC1	B6, B4
			4.11	Affluent Working Couples with Mortgages, New Homes	1.3%	ABC1	B4, B5
			4.12	Transient Workforces, Living at their Place of Work	0.3%	–	B7
	5	Well-Off Workers, Family Areas	5.13	Home Owning Family Areas	2.5%	ABC1	B4, B3
			5.14	Home Owning Family Areas, Older Children	2.6%	C1C2	B3, B5
			5.15	Families with Mortgages, Younger Children	1.9%	C1C2	B4, B3
C **RISING**	6	Affluent Urbanites, Town & City Areas	6.16	Well-Off Town & City Areas	1.1%	AB	I30, J34
			6.17	Flats & Mortgages, Singles & Young Working Couples	0.9%	ABC1	K38, C9
			6.18	Furnished Flats & Bedsits, Younger Single People	0.5%	ABC1	I32, I31
	7	Prosperous Professionals, Metropolitan Areas	7.19	Apartments, Young Professional Singles & Couples	1.4%	ABC1	I30, K38
			7.20	Gentrified Multi-Ethnic Areas	1.1%	ABC1	H29, I30

				%		
8	Better-Off Executives, Inner City Areas	8.21	Prosperous Enclaves, Highly Qualified Executives	0.9%	ABC1	I31, I32
		8.22	Academic Centres, Students & Young Professionals	0.6%	ABC1	I30, I31
		8.23	Affluent City Centre Areas, Tenements & Flats	0.7%	ABC1	K38, I31
		8.24	Partially Gentrified Multi-Ethnic Areas, Young Families	0.8%	ABC1	I31, H28
		8.25	Converted Flats & Bedsits, Single People	1.0%	–	I31, K38
D SETTLING	9 Comfortable Middle Agers, Mature Home Owning Areas	9.26	Mature Established Home Owning Areas	3.4%	ABC1	J33, B5
		9.27	Rural Areas, Mixed Occupations	3.4%	–	C10, A1
		9.28	Established Home Owning Areas	3.9%	C1	J33, B5
		9.29	Home Owning Areas, Council Tenants, Retired People	3.0%	ABC1	C8, J33
	10 Skilled Workers, Home Owning Areas	10.30	Established Home Owning Areas, Skilled Workers	4.3%	C2	C11, B3
		10.31	Home Owners in Older Properties, Younger Workers	3.2%	C1C2	C11, C9
		10.32	Home Owning Areas with Skilled Workers	3.3%	C2DE	C11, D12
E ASPIRING	11 New Home Owners, Mature Communities	11.33	Council Areas, Some New Home Owners	3.7%	C2DE	E17, E15
		11.34	Mature Home Owning Areas, Skilled Workers	3.3%	C2DE	E15, C8
		11.35	Low Rise Estates, Older Workers, New Home Owners	2.9%	C2DE	F19, E15
	12 White Collar Workers, Better-Off Multi-Ethnic Areas	12.36	Home Owning Multi-Ethnic Areas, Young Families	1.0%	C1	H29, J33
		12.37	Multi-Occupied Town Centres, Mixed Occupations	2.0%	–	C9, D12
		12.38	Multi-Ethnic Areas, White Collar Workers	1.0%	C1	H29, D12

Fig. 6.1 CACI's ACORN targeting classification (courtesy CACI)

continued overleaf

ACORN Categories	ACORN Groups		ACORN Types	Percentage of households in GB	Corresponding social grades	Corresponding 1981 ACORN Types
	13 Older People, Less Prosperous Areas	13.39	Home Owners, Small Council Flats, Single Pensioners	2.3%	C2DE	C8, C9
		13.40	Council Areas, Older People, Health Problems	2.1%	C2DE	F20, C8
	14 Council Estate Residents, Better-Off Homes	14.41	Better-Off Council Areas, New Home Owners	2.0%	C2DE	E16, E17
		14.42	Council Areas, Young Families, Some New Home Owners	2.7%	C2DE	F19, G23
		14.43	Council Areas, Young Families, Many Lone Parents	1.6%	C2DE	E16, F19
		14.44	Multi-Occupied Terraces, Multi-Ethnic Areas	0.7%	C2DE	H27, D13
		14.45	Low Rise Council Housing, Less Well-Off Families	1.8%	C2DE	F19, F20
		14.46	Council Areas, Residents with Health Problems	2.1%	C2DE	F19, F20
F STRIVING	15 Council Estate Residents, High Unemployment	15.47	Estates with High Unemployment	1.3%	DE	G22, F20
		15.48	Council Flats, Elderly People, Health Problems	1.1%	C2DE	F21, F20
		15.49	Council Flats, Very High Unemployment, Singles	1.2%	DE	F21, G22
	16 Council Estate Residents, Greatest Hardship	16.50	Council Areas, High Unemployment, Lone Parents	1.5%	DE	G23, E16
		16.51	Council Flats, Greatest Hardship, Many Lone Parents	0.9%	DE	G25, G23
	17 People in Multi-Ethnic, Low-Income Areas	17.52	Multi-Ethnic, Large Families, Overcrowding	0.5%	DE	H27, H29
		17.53	Multi-Ethnic, Severe Unemployment, Lone Parents	1.0%	DE	G22, H28
		17.54	Multi-Ethnic, High Unemployment, Overcrowding	0.3%	DE	H26, H27

– = Corresponding social grades represent national average

Fig. 6.1 continued

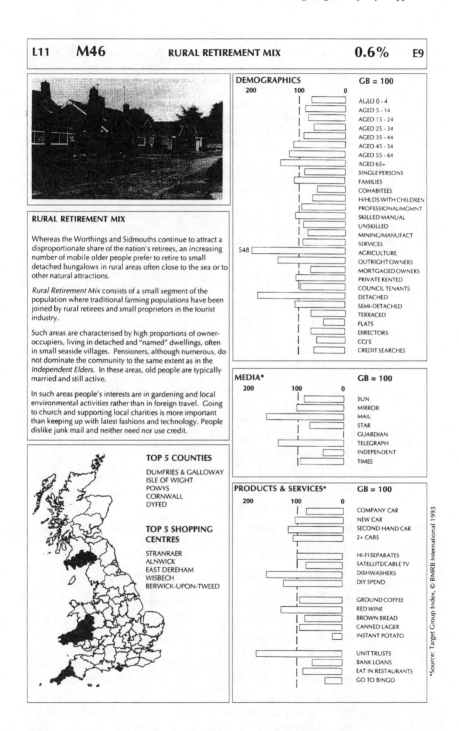

| L11 | **M46** | **RURAL RETIREMENT MIX** | **0.6%** | E9 |

DEMOGRAPHICS — GB = 100 (200, 100, 0)

AGED 0 - 4, AGED 5 - 14, AGED 15 - 24, AGED 25 - 34, AGED 35 - 44, AGED 45 - 54, AGED 55 - 64, AGED 65+, SINGLE PERSONS, FAMILIES, COHABITEES, H/HLDS WITH CHILDREN, PROFESSIONAL/MGMNT, SKILLED MANUAL, UNSKILLED, MINING/MANUFACT, SERVICES, AGRICULTURE (548), OUTRIGHT OWNERS, MORTGAGED OWNERS, PRIVATE RENTED, COUNCIL TENANTS, DETACHED, SEMI-DETACHED, TERRACED, FLATS, DIRECTORS, CCJ'S, CREDIT SEARCHES

RURAL RETIREMENT MIX

Whereas the Worthings and Sidmouths continue to attract a disproportionate share of the nation's retirees, an increasing number of mobile older people prefer to retire to small detached bungalows in rural areas often close to the sea or to other natural attractions.

Rural Retirement Mix consists of a small segment of the population where traditional farming populations have been joined by rural retirees and small proprietors in the tourist industry.

Such areas are characterised by high proportions of owner-occupiers, living in detached and "named" dwellings, often in small seaside villages. Pensioners, although numerous, do not dominate the community to the same extent as in the *Independent Elders*. In these areas, old people are typically married and still active.

In such areas people's interests are in gardening and local environmental activities rather than in foreign travel. Going to church and supporting local charities is more important than keeping up with latest fashions and technology. People dislike junk mail and neither need nor use credit.

MEDIA* — GB = 100 (200, 100, 0)

SUN, MIRROR, MAIL, STAR, GUARDIAN, TELEGRAPH, INDEPENDENT, TIMES

TOP 5 COUNTIES

DUMFRIES & GALLOWAY
ISLE OF WIGHT
POWYS
CORNWALL
DYFED

TOP 5 SHOPPING CENTRES

STRANRAER
ALNWICK
EAST DEREHAM
WISBECH
BERWICK-UPON-TWEED

PRODUCTS & SERVICES* — GB = 100 (200, 100, 0)

COMPANY CAR, NEW CAR, SECOND HAND CAR, 2+ CARS, HI-FI SEPARATES, SATELLITE/CABLE TV, DISHWASHERS, DIY SPEND, GROUND COFFEE, RED WINE, BROWN BREAD, CANNED LAGER, INSTANT POTATO, UNIT TRUSTS, BANK LOANS, EAT IN RESTAURANTS, GO TO BINGO

*Source: Target Group Index, © BMRB International 1993

Fig. 6.2 Rural retirement mix: demographic composition (*Source*: Webber 1993)

Table 6.2 MOSAIC groups and types

Type and share	No.	Descriptor	%	Typical location	Eurocode
High Income	1	Clever Capitalists	1.5	Whetstone	E1
Families	2	Rising Materialists	1.5	Wokingham	E7
(9.9%)	3	Corporate Careerists	2.4	Fleet	E1
	4	Ageing Professionals	1.7	Farnham	E1
	5	Small Time Business	2.7	Ruislip	E1
Suburban	6	Green Belt Expansion	3.4	Canvey Island	E7
Semis	7	Suburban Mock Tudor	3.2	Redhill	E2
(11.0%)	8	Pebble Dash Subtopia	4.4	Bexleyheath	E2
Blue Collar	9	Affluent Blue Collar	2.9	Kingswood	E6
Owners	10	30s Industrial Spec	3.8	Wigston	E6
(13.0%)	11	Lo-Rise Right to Buy	3.3	Long Eaton	E6
	12	Smokestack Shiftwork	3.1	Ebbw Vale	E6
Low Rise	13	Co-op Club & Colliery	3.4	Jarrow	E6
Council	14	Better Off Council	2.1	Dalkeith	E2
(14.4%)	15	Low Rise Pensioners	3.2	Barking	E2
	16	Low Rise Subsistence	3.5	Wythenshawe	E8
	17	Problem Families	2.2	Corby	E8
Council Flats	18	Families in the Sky	1.3	Peckham	E5
(6.8%)	19	Graffitied Ghettos	0.3	Motherwell	E5
	20	Small Town Industry	1.4	Hawick	E5
	21	Mid Rose Overspill	0.7	Rutherglen	E5
	22	Flats for the Aged	1.4	Clydebank	E5
	23	Inner City Towers	1.8	Southward	E5
Victorian Low	24	Bohemian Melting Pots	2.3	Tottenham	E4
Status	25	Victorian Tenements	0.1	Partick	E4
(9.4%)	26	Rootless Renters	1.5	Gateshead	E4
	27	Sweatshop Sharers	1.1	East Ham	E6
	28	Depopulated Terraces	0.8	Pontypridd	E6
	29	Rejuvenated Terraces	3.5	Southsea	E2
Town Houses	30	Bijou Homemakers	3.5	Greenford	E2
& Flats	31	Market Town Mixture	3.8	Deal	E2
(9.4%)	32	Town Centre Singles	2.1	Tewkesbury	E3
Stylish	33	Bedsits & Shop Flats	1.2	Boscombe	E1
Singles	34	Studio Singles	1.7	Surbiton	E2
(5.2%)	35	College Communal	0.5	St Andrews	E2
	36	Chattering Classes	1.9	Chiswick	E4
Independent	37	Solo Pensioners	1.9	Billingham	E2
Elders	38	High Spending Greys	1.3	Clacton-on-Sea	E10
(7.4%)	39	Aged Owner-Occupiers	2.7	Morecambe	E10
	40	Elderly in Own Flats	1.5	St John's Wood	E3
Mortgaged	41	Brand New Areas	1.0	Fleet	E7
Elders	42	Pre-Nuptial Owners	0.8	St Ives (Cambs)	E2
(6.2%)	43	Nestmaking Families	1.7	Bicester	E7
	44	Maturing Mortgages	2.7	Yate	E7

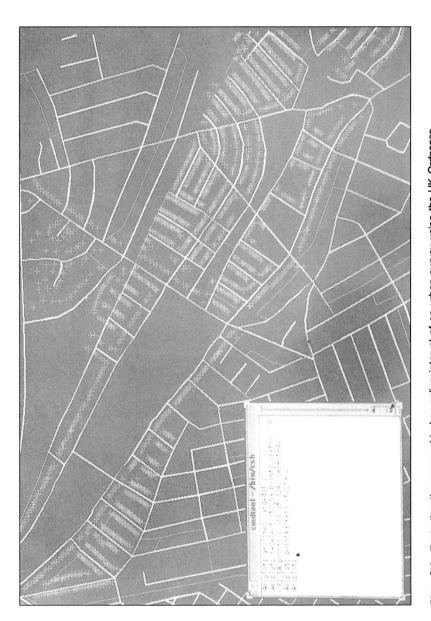

Plate 2.1 Illustrating the geographical sampling interval of an urban survey using the UK Ordnance Survey's Address Point product (courtesy: Ordnance Survey)

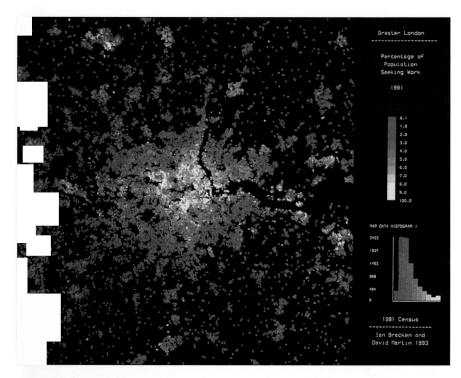

Plate 4.1 Percentage unemployment in Greater London (source: 1991 UK Census)

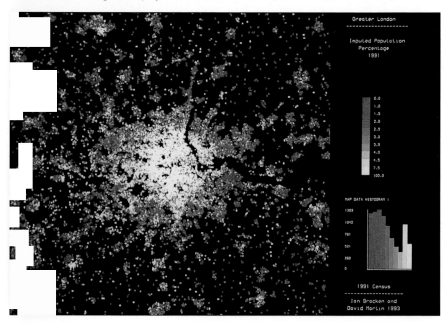

Plate 4.2 Percentage of imputed households in Greater London (source: 1991 UK Census)

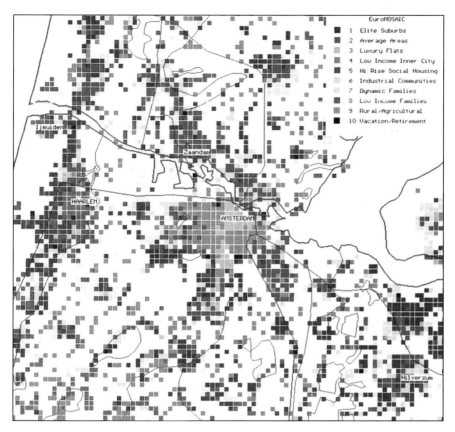

Plate 6.1 EuroMOSAIC types in Amsterdam; scale 1:400 000 (source: CCN 1993a)

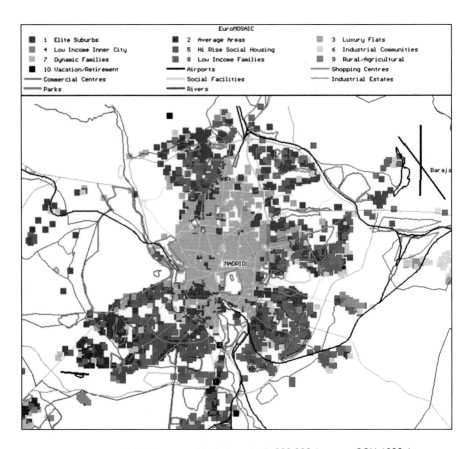

Plate 6.2 EuroMOSAIC types in Madrid; scale 1: 200 000 (source: CCN 1993a)

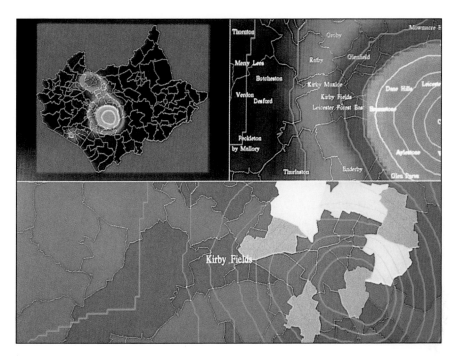

Plate 8.1 Interpolation of a population surface for Leicestershire, using the ARC/INFO proprietary GIS

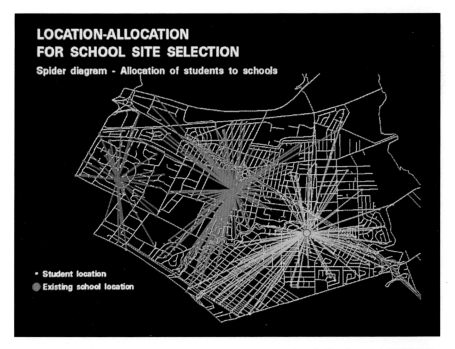

Plate 8.2 Location–allocation modelling for school site selection, California, using ARC/INFO

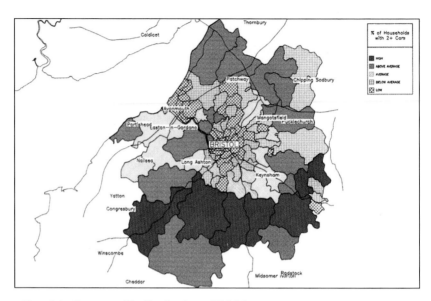

Plate 9.1 Car ownership 10 miles around Bristol

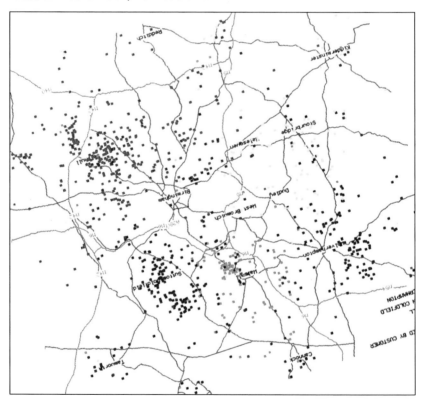

Plate 9.2 Use of point mapping to identify areas of overlap and gaps in the market, Birmingham, UK; scale 1:450 000 (source: CNN 1992)

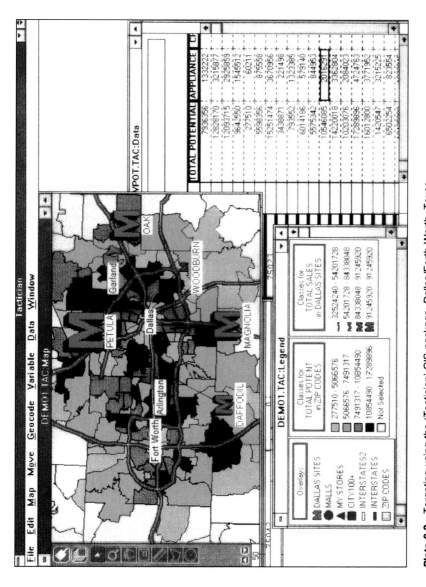

Plate 9.3 Targeting using the 'Tactician' GIS package, Dallas/Forth Worth, Texas

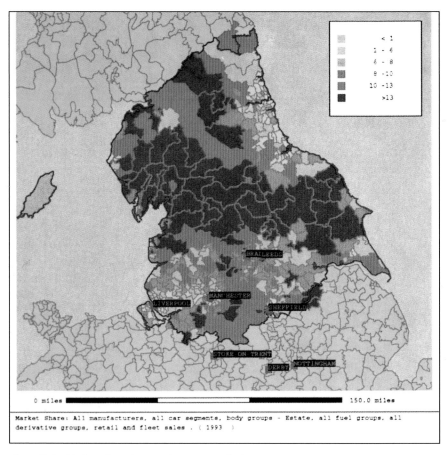

Plate 10.1 Market share for a major car manufacturer

Table 6.2 contd.

Type and share	No.	Descriptor	%	Typical location	Eurocode
Country	45	Gentrified Villages	1.5	Uckfield	E1
Dwellers	46	Rural Retirement Mix	0.6	Stranraer	E9
(7.0%)	47	Lowland Agribusiness	1.8	Diss	E9
	48	Rural Disadvantages	1.2	Evesham	E9
	49	Tied/Tenant Farmers	0.6	Stranraer	E9
	50	Upland & Small Farms	1.3	Carmarthen	E9
Institutional	51	Military Bases	0.3	Aldershot	E2
Areas	52	Non-Private Housing	0.1	Beeston	E2
(0.3%)					

Table 6.3 Comparison of 1991 census-based geodemographic classifications (after Sleight 1993b)

Classification system	Number of input variables	Non-census data used?
ACORN	79	No
PiN	49	No
FiNPiN	58	Yes FRS data
MOSAIC	87	Yes credit activity electoral roll PAF CCJs Retail access
Super Profiles	120	Yes electoral roll TGI credit data CCJs
*DEFINE	N/A	Yes credit data electoral roll unemployment stats others?
*Images	N/A	Yes NDL data others?

*Not yet complete

PAF = 'Postcode Address File'
CCJ = 'County Court Judgements'
TGI = 'Target Group Index'

there are significant differences for a reasonable number of observations, the postcode may well be assigned to a different category. There are no published details on how many postcodes share the same categories as their 'parent' EDs. One ought to note, however, that the census information is very much richer in its content than the postcode-based information. It is an open question as to whether the data are really good enough to support a reliable classification at the unit postcode level.

One other system which deserves separate mention is CCNs new 'EuroMOSAIC' product. According to CCN's product brochure:

> With the emergence of the single market, pan-European target marketing campaigns are now a reality. EuroMOSAIC, the first pan-European segmentation system, allows consistent classification of 310 million European consumers on the basis of the types of neighbourhood in which they live. (CCN 1993a)

EuroMOSAIC identifies ten different neighbourhood types, which are applied across the whole of Europe, so it is possible, for example, to compare the geographies of Stockholm, Amsterdam and Madrid (see Figure 6.3 and Plates 6.1 and 6.2).

One obvious shortcoming of EuroMOSAIC is in reconciling it with the national UK system. Referring back to Table 6.2, the EuroMOSAIC codes ('Eurocodes') for each MOSAIC type are shown in the far right-hand column. We can see that the correspondence between the higher-level MOSAIC groupings and the EuroMOSAIC groups is far from one-to-one. For example, Eurocode E2 (rather blandly labelled as 'Average Areas') ranges across seven of the eleven different MOSAIC groupings, only missing 'High Income Families', 'Blue Collar Owners', 'Council Flats' and 'Country Dwellers'.

6.2.2 Uses of geodemographics

Naturally enough, many extravagant claims have been made about the capabilities of geodemographics. A recent brochure produced by CACI in the UK identifies nine different 'applications of the ACORN Classifications for your Business':

Site analysis	Sales planning	Planning for Public Services
Media buying	Database Analysis	Market Research Sample Frames
Direct Mail	Coding	Door-to-Door Leaflet Campaigns
		(CACI 1993a)

For the analysis in this section, we will group these activities into three broader categories: market research; database marketing; and retail analysis. Let us consider each in turn.

Market research

The key capability which is offered by any geodemographics system is the facility to apply a single identifying characteristic to any neighbourhood. One

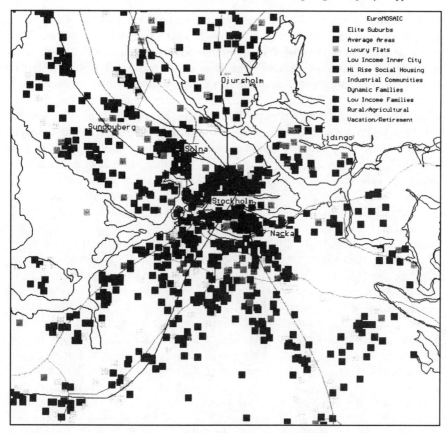

Fig. 6.3 EuroMOSAIC types in Stockholm (*Source*: CCN 1993a)

of the factors which makes geodemographics seductive is that the social geographies which are painted in this way are almost invariably extremely plausible. Potential clients might be invited to provide their home postcode, and often will be surprised to see a computer come back with a fairly accurate assessment of the type of neighbourhood in which they live.

One of the main market research applications of geodemographics comes if a client is able to produce a list of customer addresses. Each of these addresses can be postcoded, and each postcode can be assigned to a census ED. Each customer can then be ascribed the geodemographic label for the ED in which he or she resides. Once this exercise has been repeated for all of the customers in the database, it is then possible to build up a profile of the customer base in geodemographic terms.

This procedure can also be applied to market research data in order to derive more general inferences for clients who may not have customer address lists. The big initial boost to the popularity of ACORN came through its linkage with the British Market Research Bureau's 'Target Group Index' (TGI) data.

TGI records 'purchasing habits and media preferences' (Brown 1991) for a sample of around 25 000 households each year. The data cover over 400 products and 3500 brands! By using the procedure described above, it is possible to produce geodemographic profiles of respondents with different habits and preferences. An example is shown relating to the use of holiday camps in Table 6.4. The top part of the table shows the population of the UK divided into 11 ACORN groups. These have an alphabetical label (A–K) shown in column 1 of the table, and a name shown in column 2. The columns labelled 'Base' show the total numbers and percentages of population who live in areas of each type. The columns labelled 'Product' show how many people living in each of these areas have been to a holiday camp in the last three holidays, based upon grossed-up TGI data. Both base and percentage figures are presented as totals within each ACORN type (in the column labelled 'thousands') and as a percentage of the total in all ACORN groups (in the column labelled '%'). The column labelled 'penetration' shows the proportion of the population which have visited holiday camps, and is obtained by simply dividing the total number of holiday-makers by the base population. In the case of 'Agricultural Areas', 54 000 out of 1 679 000 households have visited a holiday camp, giving a penetration of 3.2 per cent. The column labelled 'Index' shows the relative product penetration and is obtained by dividing the percentage of holiday-makers who are of a particular ACORN type by the percentage of the population which is of that ACORN type. For example, 'Agricultural Areas' contain 3.8 per cent of the households in the UK, but only 2.0 per cent of the holiday campers, giving an index of 52. This shows that people living in agricultural areas are about half as likely to visit holiday camps as the average household. These indices are represented visually in the last column of the table, which shows that ACORN groups B, D and F are the ones most favourably inclined towards holiday camps. The whole exercise is repeated at a more disaggregate level for different ACORN 'types', of which there are 37, in the bottom part of Table 6.4.

We can note two assumptions which are necessary to produce analyses like those of Table 6.4 from a combination of market research data, such as TGI, and a geodemographic system, such as ACORN. The first assumption is that the sample of respondents is fully representative of the population. In TGI, for example, while 25 000 sounds like a lot of people to draw upon, we must bear in mind that in this case only 6 per cent of the sample have been to a holiday camp in the period specified. Thus the profile is based on around 1500 positive responses rather than 25 000, i.e. about thirty per ACORN type. The potential for bias is obvious. The second assumption is that because each respondent is allocated a neighbourhood category according to the census enumeration district in which he or she resides, then each ED is assumed to be perfectly homogeneous in demographic terms. For example, one would expect all persons living in a neighbourhood categorized by ACORN Type K37, 'Private houses, well-off, elderly' to be relatively old, affluent, and living in private housing. In practice this will be far from the case, as we will see in Section 6.2.3 below.

A slightly different type of market research application for geodemographics is in the design of sample frames. A typical problem here is how to select stratified samples for a survey which are fully representative of the population. Ideally, one would like to obtain a representative mixture of ages, social class groups, family status, geographical region and any number of other characteristics. The problem is that as the number of factors increases, then there are simply too many permutations to cover within the sample frame. An attractive alternative is to sample across the whole range of geodemographic types, on the grounds that these types already embrace a wide range of social and demographic characteristics. For example, much has been made of the use of ACORN to identify a pilot region for 'Charterline', a new telephone information service designed to support the UK Citizen's Charter (CACI 1993b; Randall and Furness 1993). In this study ACORN was used to profile the neighbourhoods within each county in Great Britain, and the profile compared to the national average. The objective was 'to find an area which reflected the overall profile of Great Britain as closely as possible . . . The county with the ACORN profile most similar to the national picture was Nottinghamshire and the county with the least (similarity) . . . was Inner London' (Randall and Furness 1993, pp. 4–5). Taking these results into account, the Charterline pilot scheme was launched in Nottinghamshire, Derbyshire and Leicestershire in May 1993. (See Sleight and Leventhal 1989, for more examples.)

Database marketing

Given the ability to produce geodemographic profiles for particular brands and products, as described above, then we have a potentially powerful weapon for consumer targeting. Consider again the example of Table 6.4 above. The table shows that 3.9 per cent of the population live in neighbourhoods of ACORN Type B03 – 'Cheap modern private housing'. However, between them these people accounted for 6.4% of holiday camp visits. Suppose now that a company wishes to organize a direct mailing campaign for a new or existing holiday camp. It is likely that by focusing the campaign on neighbourhoods of ACORN Type B03, it has a much higher chance of reaching potential customers than by targeting the population at random. It is this kind of process which forms the basis for using geodemographics within database marketing.

We can formalize the benefits of this type of procedure through some relatively straightforward analysis. For each ACORN group or type, we can form a 'penetration index' by comparing the target and base populations within a category. In the above example, this gives us a penetration index of $(6.4/3.9) \times 100 = 161$ for holiday camp visits within ACORN Type B03. At the other extreme, an index of only 20 is achieved within type I32 ('furnished flats, mostly single people'). These penetration indices are shown in the column labelled 'Index' in Table 6.4.

Next, we can rank the geodemographic groups in descending order on the penetration index. By plotting out the cumulative target and base populations

Table 6.4 Profile of households which use holiday camps

Household type	Product^a thousands	%	Base^b thousands	%	Penetration (%)	Index	0 50 100 150
ACORN group							
A Agricultural areas	54	2.0	1679	3.8	3.2	52	AAAAAA
B Modern family housing, higher incomes	515	19.0	6790	15.3	7.6	124	BBBBBBBBBBBBBBBBBB
C Older housing of intermediate status	527	19.4	8224	18.5	6.4	105	CCCCCCCCCCCCCCC
D Poor quality older terraced housing	159	5.9	1866	4.2	8.5	139	DDDDDDDDDDDDDDDDDDDDD
E Better-off council estates	383	14.1	5416	12.2	7.1	116	EEEEEEEEEEEEEEEE
F Less well-off council estates	336	12.4	4195	9.5	8.0	131	FFFFFFFFFFFFFFFFFF
G Poorest council estates	179	6.6	2980	6.7	6.0	98	GGGGGGGGGGGGGG
H Multiracial areas	72	2.7	1436	3.2	5.0	82	HHHHHHHHHHHH
I High-status non-family areas	80	2.9	2095	4.7	3.8	62	IIIIIIIII
J Affluent suburban housing	339	12.5	7727	17.4	4.4	72	JJJJJJJJJJ
K Better-off retirement areas	73	2.7	1942	4.4	3.8	61	KKKKKKKKKK
ACORN type							
A01 Agricultural villages	46	1.7	1263	2.8	3.6	59	AAAAAAAAA
A02 Areas of farms and smallholdings	8	0.3	416	0.9	2.0	33	AAAA
B03 Cheap modern private housing	173	6.4	1746	3.9	9.9	162	BBBBBBBBBBBBBBBBBBBBBBBB
B04 Recent private housing, young families	101	3.7	1241	2.8	8.1	132	BBBBBBBBBBBBBBBBBB
B05 Modern private housing, older children	176	6.5	2417	5.5	7.3	119	BBBBBBBBBBBBBBBB
B06 New detached houses, young families	57	2.1	1215	2.7	4.7	76	BBBBBBBBBB
B07 Military bases	9	0.3	171	0.4	5.5	89	BBBBBBBBBBBB
C08 Mixed owner-occupied and council estates	117	4.3	1668	3.8	7.0	115	CCCCCCCCCCCCCCC
C09 Small town centres and flats above shops	114	4.2	1763	4.0	6.5	106	CCCCCCCCCCCCCC
C10 Villages with non-farm employment	128	4.7	2362	5.3	5.4	88	CCCCCCCCCCC
C11 Older private housing, skilled workers	168	6.2	2431	5.5	6.9	113	CCCCCCCCCCCCCCC

Code								
D12	Unimproved terraces with old people	98	3.6	1179	2.7	8.4	136	DDDDDDDDDDDDDDDD
D13	Pre-1914 terraces, low-income families	51	1.9	550	1.2	9.3	151	DDDDDDDDDDDDDDDDDD
D14	Tenement flats lacking amenities	10	0.4	137	0.3	7.2	117	DDDDDDDDDDDDDD
E15	Council estates, well-off older workers	121	4.4	1638	3.7	7.4	120	EEEEEEEEEEEEEE
E16	New council estates	71	2.6	945	2.1	7.5	123	EEEEEEEEEEEEEE
E17	Council estates, well-off young workers	153	5.6	2032	4.6	7.5	123	EEEEEEEEEEEEEEE
E18	Small council houses, often Scottish	38	1.4	801	1.8	4.8	78	EEEEE
F19	Low-rise estates in industrial towns	166	6.1	2094	4.7	7.9	129	FFFFFFFFFFFFFFF
F20	Inter-war council estates, older people	127	4.7	1457	3.3	8.7	142	FFFFFFFFFFFFFFFFF
F21	Council housing for the elderly	43	1.6	643	1.5	6.7	109	FFFFFFFFFFFF
G22	New council estates in inner cities	46	1.7	648	1.5	7.1	116	GGGGGGGGGGGGGG
G23	Overspill estates, high unemployment	87	3.2	1362	3.1	6.4	104	GGGGGGGGGGG
G24	Council estates with overcrowding	27	1.0	584	1.3	4.7	76	GGGGGGGG
G25	Council estates with worst poverty	19	0.7	386	0.9	4.8	79	GGGGGGGG
H26	Multi-occupied terraces, poor Asians	0	0.0	81	0.2	0.0	0	
H27	Owner-occupied terraces with Asians	20	0.7	377	0.9	5.3	86	HHHHHHHHH
H28	Multi-let housing with Afro-Caribbeans	11	0.4	328	0.7	3.4	55	HHHHHH
H29	Better-off multi-ethnic areas	41	1.5	650	1.5	6.4	104	HHHHHHHHHHH
I30	High-status areas, few children	41	1.5	1050	2.4	3.9	64	IIIIIII
I31	Multi-let big old houses and flats	35	1.3	734	1.7	4.7	77	IIIIIIIII
I32	Furnished flats, mostly single people	4	0.1	311	0.7	1.2	20	IIIII
J33	Inter-war semis, white-collar workers	150	5.5	2729	6.2	5.5	90	JJJJJJJJJJJ
J34	Spacious inter-war semis, big gardens	112	4.1	2476	5.6	4.5	74	JJJJJJJJ
J35	Villages with wealthy older commuters	49	1.8	1316	3.0	3.7	61	JJJJJJJ
J36	Detached houses, exclusive suburbs	28	1.0	1205	2.7	2.3	38	JJJJ
K37	Private houses, well-off elderly	43	1.6	1248	2.8	3.5	56	KKKKKKKK
K38	Private flats with single pensioners	30	1.1	694	1.6	4.3	70	KKKKKKKKK
U39	Unclassified and unmatched respondents	0	0.0	0	0.0	0.0	0	
	Totals	2717	100.0	44349	100.0	6.1	100	

[a] Product: holiday camp within last three holidays
[b] Base: number of adults
Source: Beaumont and Inglis, 1989.

we are able to form a 'Lorenz curve' for the product in question (see Figure 6.4). With a little thought, we can appreciate that the Lorenz curve will always be convex and above the line of equality (because we always start with an index greater than 1 for the top-ranking target group, and because the gradient of the curve is equal to the index). We can also see that the steeper the curve, the greater the level of 'discrimination' that is being provided. The steepness of the curve can be measured by comparing the area under the curve with the area under the line of equality to give an index between 1 (zero discrimination) and 2 (perfect discrimination). This is generally known as the 'Target Effectiveness Index'. Finally, note that for this reason the Lorenz curve of Figure 6.4 is sometimes known simply as a 'Gains Chart'.

A key issue which we now need to address is how much discrimination is provided by the geodemographic systems in the marketplace. It would also be interesting to know whether one of the systems consistently outperforms any of the others. Not surprisingly, perhaps, commercial confidences dictate that reliable information on these issues is difficult to access. Nevertheless it is possible to make a number of assertions with reasonable confidence. One assertion is that a considerable degree of discrimination is offered by all the major geodemographic systems for a variety of products or services. As circumstantial evidence, why else would business users persist in spending good money on such products over such a long period!? There are two caveats. First, the procedure by which the Lorenz curve is produced guarantees some degree of discrimination between small areas by the use of any geographical variable. This is true because the penetration indices are ranked from high to low before the curve is plotted. It is not clear how much more we get from a geodemographic system than from a single variable such as age or social class (but see also Section 6.3 below). Secondly, the gains are typically not excessive. Openshaw (1989b) has argued that in evaluating direct mail we have to realize that a success rate of the order of 1 per cent is generally viewed as quite acceptable. Any geodemographic package which could get this up to 1.5 per cent would be excelling itself. And yet this still represents a failure rate of 98.5 per cent (Openshaw 1989b)! Note also, however, that with direct mail we are dealing with an extremely large market. Beaumont (1991) quotes an industry figure of £474 million for 1986, which has since continued to grow rapidly. In this context, marginal benefits might clearly add up to major savings.

Another assertion is that there is relatively little to choose between the various competing packages. Among the 1981 products, Super Profiles clearly has the best intellectual pedigree, having been developed over a longer period than the others and within an academic environment (Charlton *et al.* 1985; Batey and Brown, Chapter 5, this volume). This probably translates itself into a marginal, but far from decisive, performance advantage in most situations (Openshaw 1989b). Among the 1991 products, MOSAIC is being presented as a technically superior product, for example through the use of a genetic algorithm within the profiling procedure (CCN 1993b). Nevertheless, one cannot help feeling that imaginative labelling of the neighbourhood types

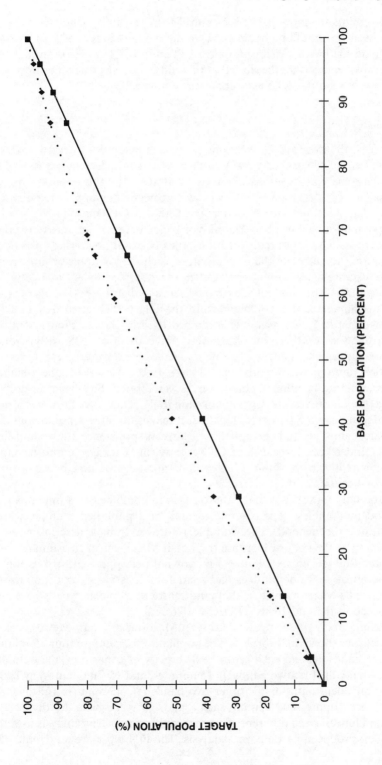

Fig. 6.4 A Lorenz curve for holiday camp visits

probably counts for more in the long run. Moreover 'it is CCN's view that a prudent decision should be based 20% on the methodology used by a system and 80% on the level of backup support behind it' (CCN 1993b). To this one would simply want to add the rider that this will cease to be true if performance variations between the systems become more noticeable.

Retail analysis

The most straightforward applications of geodemographics to retail analysis are intimately connected to GIS. We have seen from previous sections that geodemographic classifications allow us to attach a neighbourhood type to each small area in the country, such as a unit postcode or enumeration district. Suppose that we could create a coverage within a GIS showing the spatial extent, population and neighbourhood type of each area. If we were to overlay upon this coverage a set of retail locations, then basic GIS buffering operations could be used to calculate the population of each demographic group within specified distance bands around each outlet, for example 1 mile, 2 miles, and so on. In the mid-1980s, Pinpoint acquired a licence to distribute ARC/INFO (see Maguire, Chapter 8, this volume) within the UK (a licence also shared by Doric at that time), and was able to embed its PiN classification system to perform analyses of this type. Relatively naive users of GIS within retail companies appear to have been able to get excited about these types of application until quite recently (e.g. Howe 1991). However, one obvious problem with this 'banding' methodology is that straight-line distances do not reflect geographical accessibility very meaningfully. Thus if we are considering the retail potential at a particular location, it makes far more sense to consider drive time bands around an outlet or centre in preference to straight-line distances. Indeed products such as CCN's 'Environ' package comprise simple population aggregations within 15, 30 and 45-minute drive time bands around major retail centres in the country.

A marginally more convincing approach is to combine drive time analysis with geodemographics. It is possible to look at a published example which demonstrates the method. The example is drawn from a description of a method used at the Oxford Institute for Retail Management to estimate (*ante post*) potential revenues at the Meadowhall shopping centre using a combination of geodemographics and spatial analysis based on comparisons to Gateshead's Metro Centre, a shopping centre of a similar type, which had been opened in the mid-1970s (OXIRM 1990).

Geodemographic data relating to the Meadowhall and Metro Centre catchments are shown in Table 6.5. The population is split into four 15-minute drive time bands around each centre. For brevity of exposition, the data are only shown for the first two bands (0–15 minutes, and 15–30 minutes) in Table 6.5; the equivalent data for the other two bands (30–45 minutes and over 45 minutes) are suppressed. Within each band, it is possible to divide the population into thirteen different geodemographic types. This analysis uses the PiN system produced by Pinpoint Analysis. The PiN types are represented by the twelve letters (A–L) shown in column 1 of the table. Column 2 of Table 6.5 shows the number of households in the appropriate drive time band around Meadowhall which are of the given geodemographic type. Column 3 shows the equivalent figures for the Metro Centre. Column 4 shows the ratio between the number of households in each PiN type between the two centres, and column 5 gives an estimate of the amount of revenue which is generated at the Metro Centre from each PiN type within each drive time band. When we multiply columns 4 and 5 together, we can obtain an estimate of the amount of revenue which we expect Meadowhall to generate from each PiN type within each drive time band based on known performance of the Metro Centre. These estimates are shown in column 6 of Table 6.5.

Table 6.5 Market share calculations for Meadowhall (after OXIRM 1990)

(a) Overall market share calculation for 0–15 minute zone

PIN type	Meadowhall households	Metro households	Meadowhall/ Metro	Metro Sales from zone	Meadowhall sales from zone
A	402	279	1.44	300	432.26
B	0	29	0.00	00	0.00
C	19 833	26 916	0.74	37 225	27 429.17
D	4 212	12 555	0.34	12 350	4 143.23
E	51 857	62 795	0.83	25 300	20 893.10
F	11 036	19 745	0.56	9 300	5 198.01
G	25 340	12 950	1.96	9 675	18 931.62
H	19 042	9 028	2.11	3 925	8 278.67
I	4 307	10 052	0.43	2 475	1 060.47
J	3 078	1 498	2.05	0	0.00
K	5 934	431	13.77	0	0.00
L	47 569	61 689	0.77	20 875	16 096.92
Total	**192 610**	**217 967**	**0.88**	**123 325**	**108 942.61**

(b) Overall market share calculation for 15–30 minute zone

PIN type	Meadowhall households	Metro households	Meadowhall/ Metro	Metro sales from zone	Meadowhall sales from zone
A	6 058	2 681	2.26	725	1 638.21
B	0	73	0.00	0	0.00
C	52 625	41 408	1.27	14 700	18 682.08
D	27 362	31 161	0.88	14 475	12 710.28
E	51 040	84 096	0.61	9 825	5 963.04
F	35 073	33 491	1.05	7 775	8 142.26
G	56 288	22 716	2.48	4 075	10 097.45
H	25 361	27 841	0.91	1 100	1 002.99
I	2 726	4 875	0.56	0	0.00
J	311	370	0.84	0	0.00
K	1 765	1 153	1.53	0	0.00
L	95 222	93 658	1.02	6 250	6 354.37
Total	**353 831**	**343 496**	**1.03**	**59 625**	**61 418.98**

The revenue estimates produced in column 6 of Table 6.5 now need to be summed up to give an estimate of the total revenue to be generated at Meadowhall. These calculations are shown in Table 6.6(a), where only the most significant PiN types are shown for convenience. The bottom line of Table 6.6(a) shows that we can anticipate an unadjusted revenue of £670 million using this method.

Note that although most applications of this type relate to potential estimation for individual supermarkets or stores (rather than whole centres), examples are not readily available for obvious reasons of commercial confidentiality. CACI's InSite and CCN's ENVIRON are commercial packages built upon similar principles. An application of InSite within the Budgen retail group is described in a recent edition of CACI's in-house magazine *Marketing Systems Today*. The report states that

> Budgen's are planning to use the InSite system . . . for site location and marketing planning. Budgen's InSite package also includes an ACORN module which can be used to give them an instant geodemographic profile of any stores' customers.

It goes on:

> Clarence Carasco, Budgen's Development Executive, believes that 'the InSite system will help us to keep track of . . . geographical variations in demand, giving us the essential information we need to maintain a well targeted and profitable retail offer throughout . . . existing trading areas, as well as identifying locations for new opportunities'. (CACI 1993b: 13).

It is important to note two serious drawbacks with the approaches which combine geodemographics and drive times in this way. First, the assumption that catchment areas are uniform in each direction with respect to either distance or drive times is simply not adequate. A return to the Metro Centre case discussed above illustrates this point. When we look at customer flows to the Metro Centre from different directions then we find much greater levels of sales penetration to the south than to the north. Of course none of this is surprising when we consider that the metropolis of Newcastle, with many attractive alternative retail destinations, lies immediately to the north of the Metro Centre. On the other hand, shoppers approaching the Metro Centre from the south and west have few 'intervening opportunities'. Thus if we are trying to apply analogies from the Metro Centre to forecast what might happen at a new centre, such as Meadowhall, it is impossible to take a definitive view on which of these situations is likely to apply. Table 6.6(b) shows an alternative view of the potential revenues of Meadowhall which assumes that the flow patterns to the Metro Centre from south of the Tyne are more typical than those from the north. It is not clear whether this assumption is superior, not least because just as the Metro Centre is adjacent to Newcastle, so is Meadowhall adjacent to Sheffield and this will presumably affect the spatial pattern of sales in a complex way. A second problem which is highlighted by Table 6.6(b) is how to deal with the more distant drive time

Table 6.6 Retail potentials for Meadowhall (after OXIRM 1990)

(a) Using overall market share sales potential

	Catchment zone			
	0–15 minutes	15–30 minutes	30–45 minutes	45–60 minutes
PiN Type				
C	27.4	18.7	36.8	56.7
D	4.1	12.7	17.9	109.3
E	20.9	6.0	23.7	12.1
F	5.2	8.1	15.2	50.6
G	18.9	10.1	57.5	15.3
H	8.3	1.0	6.8	2.3
L	16.1	6.4	11.1	33.2
ALL	102.5	64.6	171.9	331.0

(b) Modified using south of river market share sales potential

	Catchment zone			
	0–15 minutes	15–30 minutes	30–45 minutes	45–60 minutes
PiN Type	30.1	29.4	36.8	
C	30.1	9.0	17.9	
D	10.0	9.0	17.9	
E	25.9	6.4	23.7	
F	10.2	8.8	15.2	
G	26.2	14.3	57.5	
H	9.7	0.8	6.8	
L	24.7	8.4	11.1	
ALL	138.5	78.1	171.9	*

*Not calculated; allowance of 19% for sales from outer catchment. All figures rounded to millions of pounds, mid-1988 values. PiN classification copyright Pinpoint Analysis Ltd.

bands, which we know (from elementary geometry) are going to be much bigger than the nearer bands. Table 6.6(a) implies that over 75 per cent of the revenue of Meadowhall might be drawn from customers living over 30 minutes from the centre. Experience suggests this to be unlikely, and in Table 6.6(b) a modified assumption has been applied in which customers from over 45 minutes away are assumed to provide 19 per cent of the total revenue of the centre – an exceedingly arbitrary assumption.

A third point, which is related to the first but deserves to be stressed separately, is that market penetrations cannot be assumed to be uniform within a particular drive time band. A simple example which illustrates the point is shown in Figure 6.5. This illustration is drawn from a market study undertaken by GMAP for a retail client, but has been anonymized for reasons of client confidentiality. For a group of fourteen outlets in a study region, outlet penetrations within 5-minute drive time bands were estimated by fitting a complex mathematical model to client data. An intimate relationship was found between outlet performance and penetration within the 5 and 10-minute drive time bands. Broadly speaking, the higher the penetration within the 10-minute band, the better the revenue performance of the store. In this example, both penetrations and store revenues vary by factors of more than two. Thus variations in store revenues cannot be understood simply in terms of varying retail potentials within a standard set of drive time bands: we also need to understand the factors which underpin variations in the penetration of these bands (notably the distribution of competing outlets).

6.2.3 The key issues

Despite the impressive presentation of geodemographic systems, and the apparent breadth of applications, it is vital not to lose sight of at least three fundamental methodological weaknesses in their creation. We can illustrate each of these points with reference to Table 6.7, which shows the relationship between 62 different variables used in the creation of the Super Profiles system post-1981. The column labelled 'XX' shows the average composition of the UK population on these criteria. For example, in row 3, 15.6 per cent of the population are aged between 15 and 24. The columns labelled 'A' to 'J' inclusive show indices of the population profiles of the ten top level Super Profiles clusters. For example, cluster 'G' represents 'multi-ethnic areas'. If we look at the group of country-of-birth characteristics, we find an index of 639 on 'African born', which tells us that people who live in neighbourhoods of type 'g' are 6.39 times as likely to have been born in Africa than the national average. A similar thing is true for other countries, particularly those of the New Commonwealth.

With this as background, the three weaknesses are as follows (see also Openshaw 1989b, and Flowerdew 1991):

1. The systems are cross-sectional. They relate to a particular point in time (in this case 1981) and the UK Super Profiles system was not updated between the censuses of 1981 and 1991. As we have seen above, the first profiling systems based on the 1991 UK Census data were only beginning to appear late in 1993, owing to the (entirely usual!) delays in OPCS processing of the Census data. Thus they were already two and a half years old at their launch!

2. The systems are partial in that they use census data only. The system described above is based on a classification of less than 70 census variables. It does not, therefore, take into account characteristics not recorded in the census, such as income and expenditure characteristics, favoured leisure pursuits, or interests and hobbies. Furthermore, only a partial, and to some degree subjective, subset of the available census data was used. In 1981, over 4000 different counts were available for each enumeration district.

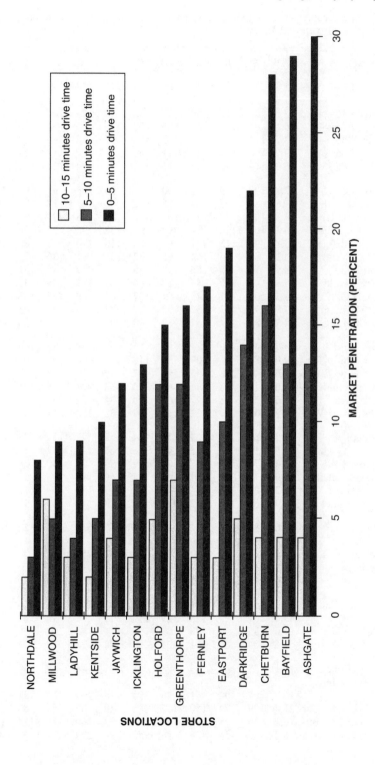

Fig. 6.5 Market share variations within drive-time bands

Table 6.7 Super Profiles characteristics: cluster means compared with national mean (from Brown and Batey 1994)

Census variable	XX[a]	A	B	C	D	E	F	G	H	I	J
Persons Aged 0–4	5.8	73	74	139	87	78	116	138	99	75	138
Persons Aged 5–14	14.0	95	60	125	97	86	115	106	87	92	138
Persons Aged 15–24	15.6	98	111	86	90	82	95	121	101	101	121
Persons Aged 25–44	26.1	91	115	135	95	90	116	100	97	80	97
Persons Aged 45–64	22.7	114	92	78	107	108	91	86	96	119	84
Persons Aged 65–74	9.7	106	109	51	111	140	73	77	118	120	61
Persons Aged 75 +	6.0	119	138	45	111	147	64	74	121	114	49
Household Size	2.7	100	79	111	100	92	106	106	91	96	117
Single Worker Hhlds	7.8	87	267	64	85	76	66	141	118	82	65
Married Couple Hhlds	44.8	100	70	136	105	105	121	78	94	87	94
2 + EA/No Ch Hhlds	23.1	104	106	114	97	95	112	96	95	98	84
Fem of Reprod Age	38.8	93	113	121	93	84	110	110	97	85	107
2 Adults + Ch 0–15	22.3	88	54	164	95	82	134	91	86	75	132
6 + Person Hhlds	3.9	79	56	66	92	52	74	217	76	101	218
Single Parent Hhlds	5.8	49	127	44	57	64	65	181	98	142	171
Lone Female OAPs	11.4	101	127	43	91	132	67	88	125	131	69
One Year Migrants	9.9	92	173	107	92	77	84	133	102	75	91
Pensioner Migrants	4.7	88	114	157	89	78	91	125	73	90	104
Rooms per Hhld 1–3	18.4	49	210	27	56	53	40	148	92	146	139
Rooms per Hhld 4–6	69.0	88	71	109	94	115	121	94	108	101	104
Rooms per Hhld 7 +	12.6	243	98	155	194	87	70	60	68	29	22
Overcrowded Hhlds	0.6	27	94	34	60	32	53	221	78	106	288
Share/Lack Bath	2.5	39	98	27	162	60	34	152	289	36	24
Share/Lack Inside WC	4.2	47	237	22	115	59	35	183	251	38	49
Owner Occupied	54.9	150	91	163	113	144	132	72	114	39	19
Council/New Town	31.6	19	44	15	43	36	60	142	52	236	276
Tied Employ/Busin	3.4	71	107	47	303	37	94	31	60	28	19
Unfurnished Rented	6.9	77	224	34	164	88	50	115	207	28	15
Furnished Rented	3.2	106	507	38	75	50	40	182	139	16	8
Second/Hol Homes	0.9	62	110	32	433	69	20	9	45	13	4
African Born	0.6	84	224	54	35	36	57	639	53	33	36
Caribbean Born	0.6	30	141	20	10	24	35	771	52	48	61
Indo/Pakistan Born	1.2	65	157	39	23	37	48	758	81	33	36
Non Commonw'lth/ EEC	1.0	163	422	71	73	54	72	163	63	37	35
Hhlds Without Car	39.0	55	120	31	56	91	67	144	131	140	151
Hhlds With 1 Car	47.4	107	93	128	109	111	125	81	90	83	79
Hhlds With 2 Cars	11.0	185	70	193	168	92	114	45	51	50	36
Hhlds With 3 + Cars	2.6	204	64	151	225	80	94	38	45	44	28
Working at Home	11.1	88	144	49	328	57	41	46	68	30	16
Travel-to-Work by Foot	16.3	57	104	41	78	95	78	107	155	123	122
Travel-to-Work by Car	21.4	129	74	150	112	115	119	65	84	84	75
Travel-to-Work by Bus	16.1	46	78	41	27	83	77	160	114	154	198
Travel-to-Work by Train	5.5	162	255	76	28	50	98	279	51	57	41

Table 6.7 contd.

Census variable	XX[a]	A	B	C	D	E	F	G	H	I	J
Employ + Managers	13.1	174	133	151	155	107	93	57	67	51	34
Professional Workers	4.1	321	186	184	89	82	80	53	47	33	20
Non-Manual Workers	31.5	135	140	135	72	121	113	96	85	81	66
Self Emp Non-Prof	5.4	80	96	78	231	97	80	82	89	59	45
Skilled Manual	19.8	39	47	69	72	98	113	108	135	136	149
Semi-Skil'd Manual	19.9	39	64	49	112	79	87	126	112	129	148
Unskilled Manual	6.1	24	48	31	61	59	75	143	135	164	198
Armed Forces	0.8	70	59	132	102	72	261	26	55	43	60
Agricultural Wkrs	3.8	28	8	36	484	29	27	4	23	19	18
Energy-Water Wkrs	3.0	62	47	89	52	81	109	46	219	113	149
Manufacturing Wkrs	25.8	63	56	94	64	100	114	115	117	129	137
Service/Dist Wkrs	52.9	132	134	111	93	106	97	96	88	88	80
Students 16+	4.3	177	137	120	103	86	91	120	73	68	74
Qualified Workers	13.3	255	191	184	102	97	83	62	57	35	22
Self Empl Workers	10.8	137	115	94	219	99	73	68	81	49	35
Part Time Workers	15.7	100	73	96	88	112	106	83	100	116	108
Male Workers	52.2	98	93	98	101	99	99	99	101	100	98
Working Wives	46.4	98	110	112	82	96	111	101	100	100	96
Male Unemployed	10.3	48	88	39	65	68	63	159	111	136	212

[a] national mean (XX) set to 100
EA = Economically Active

3. They are beset by the ecological fallacy (see Martin and Longley, Chapter 2, this volume; Martin, Chapter 4, this volume). This is perhaps the most important issue of all. Consider again the columns of Table 6.7. Cluster 'C' is labelled as 'Young Married Suburbia', and indeed there are concentrations of population aged 0–14 and 25–44 in this cluster. Yet 25 per cent of the population in this cluster are over 45. Or take another example. Cluster 'B' is labelled as 'Metro Singles', but less than 21 per cent of the households in this cluster comprise single workers (row 9 of the Table). To understand the significance of this, suppose that the UK retailer Mothercare (a chain selling products for infants and toddlers) wishes to target its customers with some new product information. The neighbourhood type 'Young Married Suburbia' might seem an obvious target. However, many people in these neighbourhoods will be neither young nor married! Even more importantly, perhaps, many others who are both young and married (with children) are located outside of cluster 'C'. Can geodemographic targeting really be the best way to reach these customers?

The most important response to these problems is to broaden the geodemographic systems to include various types of non-census data. For example, CACI's MONICA system was an attempt in the mid-1980s to use data from the electoral roll to update the ACORN product from its 1981 census base. MONICA is based on the idea that names go through phases when they are popular and normally drop out of fashion after one generation. The relationship between 13 000 different Christian names and age was examined in a cluster analysis, which classified names on the basis of their probability of being in particular age bands (Beaumont 1991, pp. 25–6). As we have seen above, the 1991 geodemographic products have tried to include a variety of non-census sources.

Nevertheless the overriding impression of the 1991 systems is their similarity to the 1981 products: it is the 1991 census data which provide the focal point. A tentative conclusion might be that geodemographics is likely to remain a popular tool only until something with greater appeal comes along. We will now consider one set of alternative methods which is just beginning to emerge.

6.3 Geolifestyles

Lifestyle databases are a relatively new phenomenon, having really only come to prominence in the UK during the 1990s. The three main providers of Lifestyle data are NDL (National Demographics and Lifestyles), CMT (Computerized Marketing Technologies) and ICD (International Communications and Data). Each obtains its information from some form of consumer questionnaire. In the case of NDL these are obtained from product registration guarantees for durable goods such as cameras and vacuum cleaners. CMT and ICD both distribute questionnaires directly to households with some form of 'free draw' incentive for completion. In this way, NDL have now collected in excess of 10 million questionnaire responses and can legitimately claim to hold information on about half of the individuals and households in the country. This list is growing at a rate in excess of 3 million per year. CMT have something like 7 million responses. ICD have tried to obtain complete coverage by linking to the electoral roll and treating its respondents as a very large sample from the entire population.

An example of one of the NDL questionnaires is shown as Figure 6.6. It can be seen that the questionnaire solicits information about the demographic structure of households, economic activity, income and credit cards, and about the preferences and hobbies of household members. A typical household data set for a household in the Isle of Wight is shown in Table 6.8.

Historically, the incentive for building large lifestyle databases of this type is for direct mail. In this case, the name and address of the respondent would also be held on each record. For a particular direct-mailing campaign, some attempt would probably be made to match the characteristics of the respondent to some desired target characteristics. Names and addresses of appropriate individuals who have the right characteristics are selected for receipt of the appropriate promotional materials, as long as they have not consciously 'opted out' of allowing their details to be used in this way. In this part of the chapter, we will be considering ways in which Lifestyle information can be used in a way which might 'traditionally' be seen as potential application areas for geodemographic technology. For these types of

A. Registration.
Please give your name and address (including postcode)
Date of purchase
Product purchased

B. Purchase
Did you receive this product as a gift?
If YES, what type of gift (birthday, wedding etc)
Where was the product purchased (e.g. department store, multiple article, TV advert. . .)
Indicate TWO factors that influenced your purchase (style, price,. . .)
Is this a product a replacement, additional product or new purchase?

C. About the customer
Gender
Date of birth
Marital status
Occupation (respondent and spouse)
Number and age of children
Income
Credit card usage
Home ownership
Time at present address
Leisure interests (select from a list of 59 activities enjoyed by self and spouse on a regular basis)
Three favourite activities (self and spouse)
Do you have a car? Make, model and year of registration

Fig. 6.6 Questions asked in the NDL Lifestyle questionnaire (Sleight 1993a, pp. 33–4)

Table 6.8 Sample of data values for Isle of Wight test data

---

postcode = PO30 5
birthmonth = 06 birthyear = 61
residence month = 03 residence year = 88
gender = 2 date of receipt = 89083
registration year = 00 registration letter = B
car model = 795 car make = 59
sun = 0 star = 0 mirror = 0 mail = 0 express = 0
today = 0 independent = 0 guardian = 0 telegraph = 0
times = 0 daily record = 0 none of these = 1
own occupation = 8 spouse occupation = 4
kids ages = 00000000/00000000/0000 marital status = 1 no kids = 1
home owner = 1 income = 06
interests =
00100000/00000010/00000010/00000011/00000000/10000001/00101001/10000000
favourite interests = 3,32,41 spouse favourite interests = 3,41,57
self employed = 0 spouse self employed = 0
has car = 1 has no car = 0 company car = 0 personal car = 1
no credit cards = 0 amex = 0 visa = 1 store = 0 cheque = 1 airline = 0

Source: NDL/GMAP.

application, the unit postcode is a sufficiently fine geographical discriminator and name and address information is not therefore shown in Table 6.8.

Five types of application will be discussed in more detail. In the first example, we will look at the use of 'raw' Lifestyle data which are aggregated by small geographical areas. In the second case, we will use 'reweighted' Lifestyle data, which have again been aggregated by small geographical areas, but have also been reweighted to make the information more representative of the population at large. In our third example we will show how lifestyle information can be used to generate small area profiles which might be either general purpose or quite specific to particular problem domains. Finally, we will consider 'psychographic' and socio-cultural applications, which try to cluster individuals rather than neighbourhoods.

It is also important to say at this point that the terms 'lifestyle' and 'psychographics' have themselves been widely discussed in the social science literature. These concepts have been in use for at least 25 years: see, for example, Alper and Gelty (1969) and Wells (1972), who give the subject explicit treatment, and Cathelat (1993, ch. 3), who offers a useful review. We do not wish to become embroiled in a discussion of definitions here, and will ask the reader to accept the *de facto* definitions of psychographics and Lifestyles that are employed here. The sources referred to above may be consulted if more rigorous definitions are required.

6.3.1 Raw data

We will illustrate this section with examples drawn from the European automotive industry. Of course this industry has always been a competitive one, with different manufacturers fighting for market share through the development of new products and through advertising campaigns which are both extensive and expensive. In recent years, the European market has become progressively more competitive with the advent of the Japanese importers, and even more recently competitive pressures have arisen from recession and the need to expand share to maintain sales levels within a contracting market.

Against this background, manufacturers are becoming increasingly aware that one way to enhance sales performance is via the optimization of the dealer network. With regard to the UK situation, twelve of the major manufacturers currently each have over 300 dealers, and the major player, Ford, has in excess of one thousand. In the majority of cases, these networks have been built up in an incremental and somewhat *ad hoc* way. Having recognized the need to improve the configuration of a network relative to the needs of the (existing or potential) market, the problem which then arises is how to achieve this. There is now a potential mechanism by which to achieve an effective model-based analysis of the market (see Birkin *et al.* 1995, for a more detailed argument, and also Clarke and Clarke, Chapter 10, this volume). In general terms, the data are also good. For example, the UK Society of Motor Manufacturers and

Traders (SMMT) produce syndicated information which can tell any manufacturer the sales penetration of each manufacturer within a postal district and by model segment. One specific problem with the data, however, is that it is very difficult to distinguish between 'retail' and 'fleet' registrations. This issue is of particular significance to Vauxhall and Ford, but is also relevant for others like Rover and BMW. The point here is that it is important to understand where the ownership of a vehicle resides, especially in order to identify the requirements for servicing, or the potential for re-marketing pre-owned vehicles. Arguably, the location of dealers is a more important issue for retail registrations than for fleet (which can be seen, for example, by comparing dealer sales patterns for small areas between fleet and retail registrations: the retail patterns are typically much more concentrated around the dealer). The problem is that the SMMT captures the place of registration of the vehicle rather than the place of residence, and interesting light can be shed on this if we compare the SMMT car PARC (PARC denotes the total number of vehicles on the road) with the 1991 census car PARC at a small area level (see Figure 6.7).

Of course the problem with the census PARC is that it tells us nothing about the make, model or age of the car, which are all crucial for marketing and network planning purposes. It is precisely this information which can be gleaned from a database such as NDL's Lifestyle Selector. The NDL information is also useful in a more conventional sense, in that it seeks to

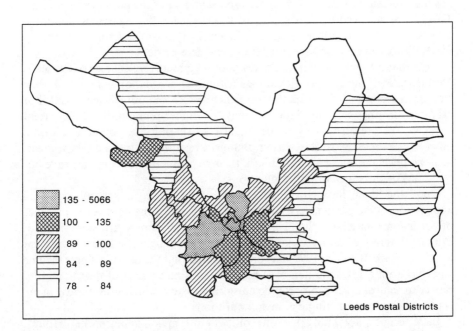

Fig. 6.7 Index of PARC vs census estimate of cars, 1991 (Leeds postal districts) (courtesy GMAP)

locate the residences of a large proportion of the customers for a particular manufacturer. This might have a use in local dealer marketing campaigns to improve the dealer's share of 'repeat business' such as parts and servicing, and re-marketing of pre-owned vehicles.

6.3.2 Reweighted data

There are two obvious problems with the use of raw Lifestyles data as outlined in Section 6.3.1 above. One is that although the databases from which these data are drawn are of ever-increasing size, each nevertheless still constitutes a sample, of which even the largest covers less than half of the households in the country. Data about individual household members are even less reliable. Secondly, because of the way in which data are collected, the databases are skewed towards particular segments of the population (in the case of NDL, towards the purchasers of durable goods; so, for example, there is a bias towards younger people rather than the elderly, and towards women rather than men).

There is an argument here that the compositional bias of the sample may not be important, because the database may be representative of the types of people which marketers typically want to reach. However, this is not seen as an altogether desirable state of affairs, as it implies that the information can only be applied to a limited set of problems. It is therefore considered desirable to reweight the data to make them more representative of the population at large. This can be achieved using a variety of sources, including aggregate published census tabulations, and the Registrar General's Population Estimates, Family Expenditure Survey and General Household Survey data.

Note that all such reweighting takes place at a relatively coarse spatial scale. Nevertheless, if the compositional bias within the population is removed, then we can expect a lot of the spatial adjustments to come out 'in the wash'. For example, in the original data we may be under-representing lone-parent families. We are also likely to be under-representing the population within inner urban areas. Most lone-parent families are clustered into just these areas, so if we reweight lone-parent families, then we will automatically improve our estimates of the inner urban populations.

Once the data have been reweighted, then there are many potential applications. One particularly powerful component of the data is the presence of an income measure. Small area variations in incomes are something of primary interest to many targeted marketing campaigns. Indeed, in a recent article Richard Webber (1992) has demonstrated that the use of a single income variable from a Lifestyle database may have a similar level of discriminatory power to that of a geodemographic clustering system which uses a great many census variables to distinguish neighbourhoods.

In a recent pilot study for one of Britain's major supermarket groups, GMAP has been developing a spatial modelling system to simulate the performance of the existing portfolio of stores and then to project the impact of changes to that portfolio. To do this we have needed estimates of the demand for products by small area. These demand estimates have been produced by a third party using a more traditional method which involves applying average expenditure rates to different demographic sub-groups by age and social class. This method will always tend to produce estimates which are too 'flat' (and for a fuller discussion of why this occurs, see Birkin and Clarke 1989). The result is that there will be a tendency to under-predict the sales of stores in affluent neighbourhoods, and vice versa. This can be corrected by applying a demand estimation methodology based on average income variation across small areas (see Figure 6.8).

Another GMAP client of longer standing is the UK bookseller W.H. Smith Ltd. Again the problem for W.H. Smith is how to combine market research on the propensity of different types of people to purchase their products, with information about the types of people who live in particular areas, in order then to produce estimates of the demand within the catchment areas of particular stores. A more direct measure of the number of people who enjoy reading books, for example, might help to refine this estimation process (see Figure 6.9).

6.3.3 Small area profiles

In the preceding section, we have already introduced the argument that there may be little to choose between the geodemographic profile of an area, and profiles based on a single Lifestyle variable such as income. However, what Webber's article apparently fails to appreciate is the additional potential of Lifestyle classifications to target multiple sub-groups at a given time. That is to say, rather than targeting different groups based on income, or on age, or on household structure (or whatever), we can use Lifestyles to target on income and age and household structure (and so on) *simultaneously*. This is possible because Lifestyle classifications record multiple characteristics for individual respondents, rather than working from area aggregates like the census or geodemographics. Thus the eternal spatial aggregation and ecological fallacy problems which bedevil geodemographic classifications are completely sidestepped.

As a simple example, suppose that we wish to market 'ContiCamp': a range of holidays offering *in situ* outdoor holiday accommodation within luxury camp sites in continental Europe. A conventional marketing campaign might begin by monitoring the response to a trial marketing campaign within one area of the country. An attempt would then be made to link respondents to the campaign with their place of residence, and hence to a geodemographic code for that neighbourhood (see Martin and Longley, Chapter 2, this volume). From an analysis of neighbourhood types, a ranking of potential response rates from particular neighbourhood types could be established, and used as a basis for targeting future mailings. Within a Lifestyle-based classifier, on the other hand, we can begin by identifying specific target groups for our new

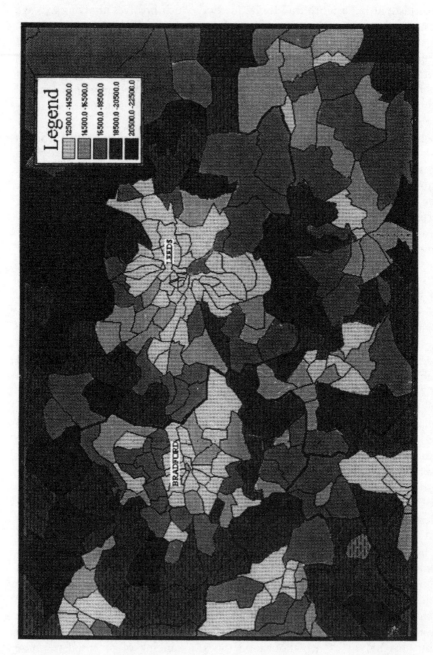

Fig. 6.8 West Yorkshire income distributions (courtesy GMAP)

Fig. 6.9 Propensity to read books in the West Midlands (courtesy GMAP)

product: for example, let us look at families which are not too old and have an interest in both continental holidays and camping. An example profile is shown in Figure 6.10. In this case it is one which might be of use in regional or television advertising, but similar profiles could be produced at smaller spatial scales.

6.3.4 Psychographics

In Section 6.2 we described the way in which small area population data allow geodemographic classifications to be produced by clustering together neighbourhoods of a similar type. For any small area, such as a UK enumeration district (ED), there are two implications. One is that all of the households in the ED are assumed to be associated with a particular neighbourhood type (and recall that there are two dubious features of this assumption: first that all the households are the 'same', and second that an administratively defined ED can be viewed as some sort of identifiable neighbourhood). The second is that we can view this neighbourhood as similar to other EDs in other parts of the UK, or even elsewhere in Europe.

Using Lifestyle data it is appropriate to attempt a similar type of clustering of respondents. In this case we can compare individual or household characteristics and identify groups of individuals or households who may be expected to share interests, attitudes, and patterns of consumption patterns. Such systems are typically referred to as 'psychographic' classifications: a well-known example based on CMT's Lifestyle data is 'Persona'.

A particularly glossy application of the psychographics concept is the 'Niches' system marketed by Polk Direct in the United States (Polk Direct 1993). In this system, 26 different clusters (niches) are identified, and each is coded under a different letter of the alphabet, from 'Already Affluent' to 'Zero Mobility'. Furthermore, the letter codes are ranked by both income and age. Thus niches A–E have the highest household incomes, while niches T–Z are the least affluent. Within the groups, niche A has the lowest average age, and niche E the highest. Similarly, niche T has the youngest members amongst the least affluent niches, and niche Z the oldest (see Table 6.9). The marketing applications of niches appear to be very similar to those of geodemographics: for example, the niches customer profile report shown in Table 6.10 is exactly analogous to the geodemographic profile of Table 6.4. Each niche also has its equivalent of the geodemographic pen portrait, comprising a description of a typical niche member (see Figure 6.11).

One way in which the geography can be brought into psychographic systems is simply by profiling zones on the basis of their household types. Figure 6.12 shows an example of this in which census tracts in the city of Logan, Utah, have been profiled by two contrasting niche types, F and S. Clearly there is considerable potential for describing geographical variations in terms of the relative concentrations of 26 different niches in this way.

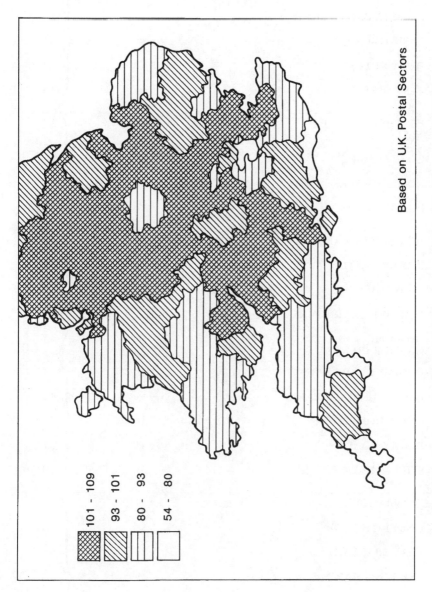

Based on U.K. Postal Sectors

101 - 109	
93 - 101	
80 - 93	
54 - 80	

Fig. 6.10 Customer likelihood for European camping holidays (courtesy GMAP)

Table 6.9 Niche profiles (*Source*: Polk Direct)

Niche	Average house-hold incomes	Average age of head	Approximate % of niches households	
A. ALREADY AFFLUENT	Over $75 000	30		0.8
B. BIG SPENDER PARENTS		44		4.5
C. CASH-TO-CARRY		50	15	4.1
D. DIAMONDS-TO-GO		52		3.3
E. EASY STREET		65		2.0
F. FEATHERING-THE-NEST	$50 000–75 000	34		2.3
G. GO-GO FAMILIES		44		1.1
H. HOME HOPPERS		47	13	5.7
I. IRA SPENDERS		69		4.1
J. JUST SAILING ALONG	$30 000–49 999	33		4.6
K. KIDDIE KASTLES		42		6.2
L. LOOSE CHANGE		44	20	1.9
M. MID-LIFE MUNCHKINS		53		3.6
N. NOMADIC GRANDPARENTS		69		4.1
O. OODLES OF OFFSPRING	$20 000–29 999	28		3.2
P. PARENTUS SINGULARIS		38		4.3
Q. QUIET HOMEBODIES		44	20	5.3
R. ROCKY ROAD		45		5.8
S. STILL GOING STRONG		65		1.3
T. TOTEBAGGERS	Under $20 000	29		2.3
U. UNDER-THE-CAR		33		1.5
V. VERY SPARTAN		37		9.6
W. WORKING HARD		49	32	2.4
X. X-TRA NEEDY		69		5.2
Y. YOUNG-AT-HEART		70		3.2
Z. ZERO MOBILITY		71		7.6

Demographic profile	Product interests
White collar, few kids, high home value	Stocks, home improvements, import cars, extensive travel, multiple credit cards
Traditional families with kids, white collar	Video cameras, home computers, camping equipment, home improvements
High home ownership, 2 or more adults, no kids, high education	Stocks/bonds, apparel, home remodeling
Two or more adults, 1 or more kids, highly mail responsive	Multiple new car buyers, catalog by phone, frequent flyers
High home ownership, white collar, 2 + adults	Home computers, high credit card spending

High presence of kids, white collar, high mobility	Kid and baby items, video cameras, catalog buyers, multiple credit cards, import car loyalty
High presence of kids, high mobility	Home computers, camping equipment, kids' items, multicar owners, jogging apparel
Low home ownership, low presence of kids, high mobility	Racquetball, home furnishings
High home ownership, 2 + adults	Catalog by phone, stocks/bonds, multiple credit cards

Low presence of kids, high mobility, renters college education	Camping equipment, domestic business travel, beer
Large households, 2 + adults/2 + children, high home ownership	Multiple new car buyers, kids' clothing, computers, multiple credit cards
High home ownership, low presence of kids, 2-person households	Import loyal new car buyers, stocks, domestic business travel, video cameras
High home ownership, high presence of kids	Home improvements, tapes and records, children's items, multiple credit cards
Two persons, grandchildren, mail responsive	Bonds, gardening, foreign travel

High mobility, very high presence of children	Import car loyalty, kids' clothing, baby items, tape rentals
Single parents, not mail responsive	Automotive tools, kids' clothing, food extenders/helpers
High home ownership, low home value	Camping equipment, home improvements, tapes and records
Blue collar, low home value, low presence of kids	Used car buyers, camping equipment, food extenders/helpers
Highly mail responsive, grandchildren	Luggage, camping equipment, baby items, domestic vacations, multiple credit cards

Mostly singles, female headed, high mobility	Men's and women's apparel, biking gear, sporting goods, high credit card ownership
High mobility, high home ownership	Automotive tools, jewelry
Renters, low presence of children, blue collar	Low credit card and car ownership, beer cigarettes/cigars
Blue collar, low education, mostly female heads	Kids' clothing, cigarettes/cigars, domestic car loyalty
Low home ownership, female heads, low education, grandchildren	Food extenders/helpers, low credit card and car ownership
Two or more adults, highly mail responsive, low mobility	Luggage, travel, own only one car but bought new
Two adults, low mobility, low education	Single car owners, auto service centres, diet control

Table 6.10

(a) Major applications of niches

The marketing applications of Niches are virtually unlimited. Here are some key ones:

- Customer Profiles (as shown below, section b)
- Direct Marketing
- Mail and Telemarketing Audiences
- Print and Broadcast Section
- Creative Targeting
- Product Positioning
- Analysis of Market Potential

- Customer & Other Databases
- Cross-selling
- Survey Analysis
- New Product Development
- Site Location
- Research Sampling
- Strategic Planning

(b) Sample Niches customer profile report

Rank		Niche	Number of hhlds	Percent of hhlds	Number of customers	Percent of customers	Niche penetration %	Index
1	G	Go-Go Families	880 000	1.1	5 500	5.5	0.625	500
12	E	Easy Street	1 600 000	2.0	2400	2.4	0.150	120
26	Z	Zero Mobility	6 080 000	7.6	800	0.8	0.013	10
Totals			80 000 000	100.0	100 000	100.0	0.125	100

HOME HOPPERS

Being Janet's mother has never been easy. It's hard just keeping track of her. She and her husband are never home and they've *moved* so often. I wish they would just settle down and buy a house, but they prefer apartments because his job involves lots of traveling. Needless to say, with all their different places, they've accumulated a lot of furniture. Every time they move, Janet worries that the movers aren't going to be careful with her antique coffee table, or her new couch, or her new bedroom set – you get the picture, I'm sure.

They're always out playing racquetball or doing something else to keep fit. So I don't talk to them much and there aren't any grandkids. But I do get to visit a lot and it's never boring. There's always so much new to see and my friends love to hear about it when I get home.

| Avg. Income $50,000 – $75,000 | **H** | Average Age 47 |

Fig. 6.11 An example of a Niches pen portrait (*Source*: Polk Direct 1993)

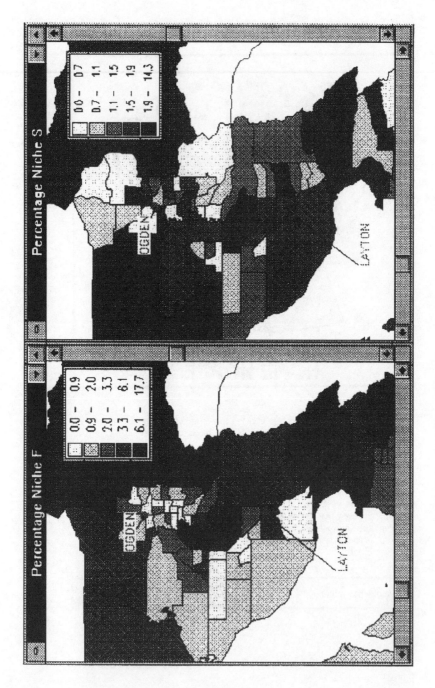

Fig. 6.12 Contrasting Niche concentrations in Logan, Utah

There are two significant differences between the geodemographic and psychographic approaches. First, the different niches are geographically dispersed and not concentrated into neighbourhoods like ACORN or MOSAIC groups. Secondly, while geodemographic clusters are assembled using data which are fairly comprehensive, lifestyle data are inevitably partial, relating only to individuals who have responded to a particular survey or questionnaire. The dynamic nature of lifestyle questionnaires also poses certain problems. If new occupiers move into a house, for example, the classification of the household will only change if and when the new residents fill in the appropriate questionnaire. One would imagine that the effects of household mobility will become more problematic as the lifestyle databases mature. It is also important to note that some of the aggregation difficulties which apply to geodemographic systems are also applicable to psychographics. Each of the psychographic clusters is likely to conceal much diversity in the behaviour of the individual households, just as the households within a geodemographic neighbourhood are, in reality, a diverse rather than a homogeneous mixture.

6.3.5 Socio-styles

It is also worth noting that the applications which we have described in this section are all drawn from either the UK or the US. This is largely because stronger confidentiality laws within Europe have tended to preclude the generation of databases based on individual responses to lifestyle questionnaires. One important exception to this is the 'Socio-styles' system developed by French researchers in the 1970s, and subsequently applied across Europe and in the United States. The review presented here is based on Cathelat (1993), to which interested readers are directed for further details.

The method is based on a relatively limited range of questionnaires which combine a variety of information types, including behaviour, attitudes, emotions, and cultural values. The pan-European Euro-Socio-Styles system is based on 24 000 questionnaires distributed across 15 countries (a response frequency of around 1 in 15 000!). The types of questions asked are much broader than the questions associated with the lifestyles databases. For example, one question shows 14 people dressed in different sets of clothing and asks the respondent 'In an ideal world . . . which type of clothing would you like to wear?' (Cathelat 1993, p. 128). The responses are subjected to complex analyses, which produce something like the factor scores of a conventional factor analysis. In order to present socio-styles information visually, it is possible to focus on the two major factors. In the generic system, the two major factors are labelled as 'values' and 'mobility'. The Euro-Socio-Styles 'map' is shown in Figure 6.13, where we can see 16 clusters of like-minded individuals. In the 'south-east' of the map (Cathelat's terminology) we find settled individuals with strong moral values, labelled 'strict', 'citizen' and 'gentry'. At the other (north-west) extreme, we find 'dandy', 'business' and 'rocky', who are

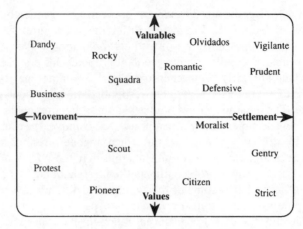

Fig. 6.13   The basic socio-styles map for the European market (*Source*: Cathelat 1993)

mobile and seek 'valuables' (i.e. wealth!) rather than values. The two extreme types in the socio-styles equivalent of pen portraits are described as:

Dandy – hedonist youth with modest income seeking welfare structures
Strict – overly repressed puritans in favour of social control.

Applications of socio-styles are based largely on panel surveys where the purchasing behaviour of individuals is monitored and tagged to the socio-styles characteristics of the responding individual. An example is shown in Figure 6.14, which presents a contrast between the users of washing powder and liquid detergent for a particular brand (these examples are overlaid on the French socio-styles mentalities, so broadly 'moralizer' equates to 'strict', and 'profiteer' to 'dandy'). There is a clear contrast between the different groups which, it is argued, is much less distinct for other brands (Cathelat 1993, ch. 9).

A second example is shown as Figure 6.15, where media interests in Norway and Italy are superimposed on the same base to show, for example, that the *Famiglia Cristiana* magazine in Italy reaches the same types of people as Norway's *Sunnmorsposten*. In an age which has seen continued expansion and internationalization of media interests there are clearly potential insights here for media moguls! Additional applications are seen as defining target populations for products, improving targeting techniques, identifying the best methods for promoting new or existing products, and providing the kind of market information which can support these applications. Clearly the socio-styles approach presents some interesting insights here which go some way beyond geodemographic and lifestyle approaches for some types of application, with a more particular emphasis on brand positioning and advertising rather than on direct marketing and spatial analysis.

(a) Buyers of brand A washing powder

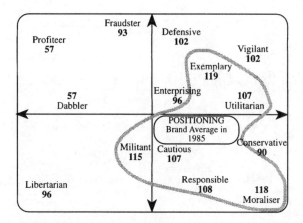

(b) Buyers of brand A liquid detergent

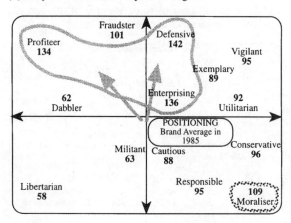

Fig. 6.14   'Socio-style' comparison of washing powder and liquid detergent users
(*Source*: Cathelat 1993)

6.3.6 Future developments

There are two more speculative issues which deserve brief mention. The first is
that although all the examples which we have presented are based on spatial
areas no smaller than postal sectors, it is quite possible to repeat the exercise at
a finer spatial scale. Indeed, Equifax claim to have created a profiling system
from lifestyles data based on the unit postcode. The second and related point is
that it is quite feasible to think about creating neighbourhood clustering
systems based on Lifestyle data rather than the traditional census data. Given

(a) Competitive equilibrium map showing the position
     of news magazines in Italy

(b) Analysis of competitive positioning of the media
     (daily newspapers in Norway)

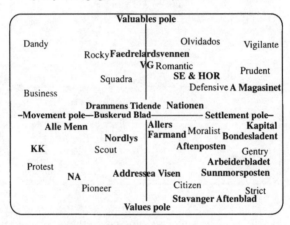

Fig. 6.15 'Socio-style' analysis of magazine markets: Norway and Italy compared
(*Source*: Cathelat 1993)

the similarity in performance levels between existing geodemographic systems,
there is probably no good reason to expect lifestyle-based clusters to perform
much differently. One potential benefit is that such systems might be much
cheaper given freedom from OPCS royalty charges (and see also Openshaw
1993a, for more speculation along these lines). Furthermore, the classifications
might be updatable on a far more frequent basis than those based on the
decennial census.

6.4 Comparison of geodemographics and geolifestyles

As I have argued in Section 6.2, there are many flaws to geodemographics. The principal redeeming features are that the technology is widely understood and accepted; is reasonably cheap, robust; and is known to work, at least to a degree. By comparison, lifestyles have some obvious advantages. The data are being continually replenished, so the systems are much more updatable. Next, information is collected on non-census variables like incomes, lifestyles and behavioural data. This means the systems are potentially more relevant to marketing. Furthermore, there are technical advantages through the use of individual rather than aggregate data, and the systems should be even cheaper. A difficult factor to evaluate is whether boredom with established methods might have a role to play! The obvious disadvantages with lifestyle databases are that they are known to be unbalanced in their coverage of the population; and that the technology is not so well proven or understood. We simply cannot be sure yet that lifestyle classifiers will work as well as geodemographic systems, let alone better.

7

Marketing spatial analysis: a review of prospects and technologies relevant to marketing

Stan Openshaw

7.1 Introduction

Seemingly, few organizations have as yet made much money out of applying sophisticated rather than simple spatial analysis technology in a marketing context. So far the major financial returns from the commodification of geography have mainly come from the sale of digital map products, and much smaller amounts from mathematical modelling of spatial interaction data in retail decision support, from the application of conventional aspatial statistical methods to census and client data, and from the reorganization of space via redistricting and location–allocation modelling. This is perhaps surprising, in view of Openshaw and Goddard's (1987) earlier comments on the prospects for enhanced spatial analysis founded upon the vast range of diverse databases that are now available in readily accessible form (see also Martin and Longley, Chapter 2, this volume). It also suggests that there is still plenty of scope, given the right opportunities, for the further commercial exploitation of the quantitative geography tool-box.

This chapter seeks to identify and develop spatial analysis technology that is, or could be, of interest in marketing contexts. First, though, Section 7.2 provides a review of expectations and perceptions from both a geographical and a marketing perspective. Section 7.3 then focuses on some of the key changes in information technology (IT) infrastructure relevant to marketeers, focusing in particular on the new opportunities this creates for applied spatial analysis. Section 7.4 briefly reviews some of the more marketing-relevant methods and Section 7.5 outlines some possible future developments.

7.2 Views, perceptions and expectations

7.2.1 Geographers and marketing

To many quantitative geographers, marketing geography is associated with so-called 'retail geography', and simplistic notions of market analysis based around a single central business district (but see Guy 1994). Most geographers seem never to have really understood commercial needs or appreciated the potential applied opportunities for their technology in a marketing context, or indeed to have displayed concern for the needs of applied users. As a result, in common with the rest of social science, there has never been any real stimulus into developing commercially relevant variants of it. Indeed, even in those areas where the applied and commercial applications are now so glaringly obvious, it seems to take at least a decade for the linkages to be made and the technology-transfer process to be started. It would appear, on reflection, that this is as much related to the life-cycle of the academics involved as to the commercial potential of the techniques. Indeed, 'nasty' profit-making commercial activities have traditionally happened, if they happened at all, near the end of an academic's active research career. This is not surprising when academic performance and career prospects have been and still are measured in terms of academic papers published and by research grant awards, rather than by pounds sterling or dollars generated via commercialization, product royalties, and licensing agreements. All too often the academic peer groups which judge performance wrongly assume that making money is easy, facile, and that it is not scholarship: in short, such activities are reduced to the 'anyone could do it if they wanted to' category. Reality is, of course, quite different. Very, very few academics have any of the necessary commercial skills even to get started, let alone to be successful.

Geographers have been particularly slow to appreciate the commercial relevance and monetary value of many basic geographical analysis and modelling skills. Moreover, the GIS software industry arguably reinforces this trend by not providing much support for real spatial analysis functionality, seemingly on the spurious grounds that there is no current market demand for it: for example, such spatial interaction and location–allocation facilities as are generally available are of only extremely limited practical value (but see Maguire, Chapter 8, this volume, for a different view!). From a marketing perspective, the principal attraction of spatial analysis is still probably a psychological one. Marketeers seem to feel that geography is important in that they know that there are major geographic variations in the demand for products. Maybe they feel that geographers should be able to help them perform better and that there might be methods that geographers know about that could be beneficial to them. Those in the industry who believe this will probably already be displaying more confidence in the value of geography than do many geographers!

At the same time the main problem that geographers face is knowing what might be needed in a marketing context. Seemingly few geographers know much or anything about the needs of marketeers and they are thus unable to be of much assistance even if they wished to be. Getting started is another problem. Yet in principle any geographer reasonably proficient at stepwise regression, logistic regression, and discriminant and cluster analysis already knows much of the basic marketing-relevant technology. This knowledge, together with a PC, a statistical package, a London address, a persuasive demeanour, and a little luck in making initial contacts and networking, is probably sufficient to create a small marketing consultancy. A few unique selling points and the ability to hype-up mediocre results through mild exaggeration, might also be regarded as useful. Particularly attractive at present is an ability to use neural-net-based methods, whilst waiting in the wings is the next wave of trendy technology based on genetic algorithms, which is likely to create a major bio-marketing industry sometime soon. However, it is arguable as to whether there is much or any need to be ahead of the field in terms of technical sophistication. Academic arrogance tends to be a factor here. Maybe the market does not need the highest quality and most advanced technology, but these factors are precisely what drive many academic geographers to aspire to excellence and simultaneously to diminish their commercial prospects.

To summarize, it would seem that many spatial analysts, because of their knowledge of multivariate statistical methods and latterly in GIS, possess most of the basic and essential analysis skills needed to be successful in marketing. Indeed when such people join marketing consultancies they often do well, or at least appear to. However, if they remain as academic geographers and are deprived of both contact networks and key marketing databases, they are not likely to make much, if any progress, except as a limited source of consultancy advice and ideas for others to exploit. Yet at the same time it is evident that there are mutual benefits to both the marketing industry and to geography from closer collaboration. The geographer might gain access to data not in the public domain, new publishing opportunities may arise and there is at least some prospect for technology transfer and commercialization. The marketing industry might gain access to a largely untapped skill base. The question is, however, which methods, which applications, and which new products might be created through such collaborations? It is also pertinent to ask whether there is an explicitly spatial focus in any such collaboration, and whether it might be founded upon explicitly geographical analysis rather than on applied statistical analysis. These are hard but fundamental questions to ask a geographer, especially when, as a profession, they have in general been extremely slow and possibly reluctant to develop applied spatial analysis skills. If lack of awareness is the main problem then maybe the comments in this chapter may help.

7.2.2 Whither applied spatial analysis?

If little applied use is made of spatial analysis anywhere, why should marketeers be interested in being among the first to use it? There are a number of possible responses to this question, which can be reduced to five key factors:

1. A perceived opportunity to gain competitive advantage by commercial exploitation of potentially useful methods of analysis and modelling of spatial data.
2. Better processing of spatial data, thus constituting an information industry that can generate wealth by adding value.
3. Concern that others might be gaining a commercial advantage by being first to appreciate the opportunities of spatial analysis.
4. Concern about the lack of basic academic research in many of the core areas of marketing and a feeling that geographers could or should be interested.
5. There is some evidence of usefulness – for example, the retail modelling activities of GMAP (see Clarke and Clarke, Chapter 10, this volume).

Another related factor is the rapid rate of change in many of the important aspects of the marketing world, although seemingly not all of these developments are visible to insiders. In general, though, it is clear that the world in which marketeers operate is becoming ever more complex: there are more competitors and profit margins are under pressure, but at the same time there is much greater data availability, much faster and cheaper computer processing, and new and more sophisticated processing technologies. The dilemma is whether these technological developments can either be safely ignored or, alternatively, safely exploited! It is in these turbulent conditions that the geographer has to try to establish his/her credentials.

Perhaps most important of all, there is a growing need for an awareness of what is going on and where the technology is leading. All of this takes place at a time when competitive tensions are increasing, with a growing concern about whether or not apparently new, better, and much improved methods and technology might have emerged. Maybe, if you understood what was happening and knew something about the nature of the new technologies, you might be able to do better in the applications areas relevant to your business concerns. As many concern geographical things, then geographers may have something to offer here.

7.2.3 A geographical opportunity

It is argued then that there is an emerging opportunity to use geographical information and to apply spatial analysis tools in marketing and that for human geographers, at least, there is the prospect of an interesting new area of academic study. In seeking to build this partnership it is important to be aware of the potential constraints on both parties and to appreciate the institutional barriers that may exist and which can so easily get in the way of the research (but see Clarke *et al.*, Chapter 12, this volume). However, it is also useful to consider the changes that are currently under way in IT which may well be regarded as a major driving force in making spatial analysis a viable technology in a marketing context.

7.3 Significant changes in marketing IT

There are three key processes of change that are relevant here:

1. computing hardware and software;
2. databases; and
3. analysis technology

7.3.1 Computing hardware

The changes in computing hardware are perhaps the most obvious and notable. Everyone knows that computers are becoming faster, smaller, and cheaper; but the rate of change is now quite spectacular. It is perhaps of only passing historic interest to note that in 1970 a large IBM mainframe (the largest model 370/165) came with 3 megabytes of RAM and each massive disk drive had a 100-megabytes capacity. Twenty years later most PCs would outperform this monster for about 10 000 times less money in real terms. The improvement in cost performance has been amazing. For example, a large Amdahl mainframe (model 5860) would have cost at least £2 million in 1985 and was rated at about 3.1 LINPACK megaflops (million floating point operations per second); at least five times faster than the IBM 370/165 could manage in 1970. Yet by 1990, a Sun Sparc 1 work-station costing about 200 times less in real terms was rated about half the Amdahl's speed. In 1992, the latest Sun work-stations were running about seven times faster, had a bigger RAM, and could handle equivalent or more disks. This Sun work-station was advertised as the equivalent of a Cray I supercomputer which in 1975 would have cost 20 million dollars.

These changes in price performance are set to continue; indeed at the top end even more dramatic changes are anticipated. Many computer scientists now talk about an impending new era in computation due to the emergence of massively parallel computing hardware; see Daedalus (1992) for a review. It now seems likely that during the 1990s the economics of information processing will undergo a major revolution. In 1993, the UK's fastest supercomputer, the Cray ymp8 at the Rutherford Atlas Laboratories, is rated at a peak speed of 2667 megaflops. It has 8 processors. Most users will probably be lucky to gain much more than a factor of 10 speed-up compared with a Sun Supersparc work-station. Currently, the world's fastest super-computer is rated at about 131 000 megaflops and has 1024 processors (Dongarra *et al.* 1993). Users would be lucky to get half or less of the peak speed but this peak is nevertheless several thousand times faster than a Sun work-station. By the late 1990s it is confidently expected that peak speeds in excess of a teraflop will be commonplace (1 teraflop = 1 000 000 megaflops). The implication for marketing is that fairly soon, indeed if not now, affordable hardware will exist which is able to handle any conceivable present or future marketing-relevant database and analysis problems. As the leading hardware developments enter teraflop speeds, so the rest of us will be left with cheap gigaflop (1000 megaflop) hardware; if indeed it is possible to create marketing applications big enough to need it!

However, there are at least three areas where low gigaflop speeds might be useful now: retail modelling, credit scoring, and database marketing. As hardware costs fall during the next few years and gigaflop work-stations appear, so the number and range of marketing applications will dramatically increase. Targeting of clients and modelling predictions of their needs for consumer products will reach unparalleled degrees of sophistication and accuracy. This claim rests on the view that for many commercial applications it is possible to trade memory and computer speed for lack of knowledge engineering, and even software effort. Waltz (1990) writes:

> Today's massively parallel machines present [artificial intelligence] AI with a golden opportunity to make an impact, especially in the world of commercial applications. The most striking near-term opportunity is in the marriage of research on very large databases with case based and memory based AI. (p. 1)

Currently, there are few, if any, marketing applications of this technology. Faster computers do not merely speed up current applications but they create major new opportunities for new products and new approaches. This is as true in the academic world as it is in the commercial world.

As computer speeds increase and costs diminish, so the economics and capabilities of information processing relevant to marketing will be revolutionized. New forms of analysis and modelling will become practicable commercial propositions and, perhaps more importantly, spatial analysis technologies might well become a mainstay of a new computational marketing era. As marketing databases become more sophisticated and customer information systems become operational and mature, so important new opportunities and needs for spatial analysis and modelling will emerge.

It is interesting to note how the economics of geodemographic processing have already changed dramatically in some areas of marketing. The 1981 UK Census ACORN system (A Classification of Residential Neighbourhoods: see Batey and Brown, Chapter 5, this volume) was allegedly based on a minicomputer. The development of the 1981 Super Profile system consumed the equivalent of £2m of computer time on a large mainframe and a single run took about two days of CPU time (Charlton *et al.* 1985). Today a 1981 Super Profile system could be built on a PC in about two days, for about 1/20 000th of the 1983 cost. No wonder Openshaw (1993b) can claim that most large users should today have their own customized geodemographic systems: indeed, there is no good reason why they should not have several, each tuned to different market sectors. Today the principal cost is the census data, not the computer processing. On the other hand, who wants to create in the mid-1990s a geodemographic system no better than was possible in 1983? Maybe the extra computing capability needs to be invested in building significantly better geodemographic systems. A much better strategy here as in other areas of marketing involves finding better ways of using unlimited and cheap computing power to improve what you do, not just to make it cheaper. This is discussed further in Section 7.4; indeed, the development of better targeting tools is probably the single most important present-day application area for spatial analysis in marketing.

7.3.2 Databases

The last few years have seen the increased use of large marketing databases. In the UK, database vendors or brokers such as NDL have databases containing details of over 14 million people and CMT of over 7 million (see Birkin, Chapter 6, this volume). ICD, Equifax, CCN, GUS, and Littlewoods have some details about nearly everyone aged 18 and over, thanks to computer copies of telephone directories, registers of electors, and other public information such as shares registers and County Court judgements for bad debt. Very large-scale sample surveys also provide details of the lifestyle and behaviour patterns of millions of people. These have been supplemented by other survey forms distributed with guarantee cards, by sponsored shopper surveys, and by estimates of key variables such as income, age, and credit worthiness. In addition, many large organizations hold comprehensive customer information databases and have often spent millions building usable systems.

By comparison, the 1991 UK Census and the SAR look much less attractive. In a marketing context, accurate census data for April 1991 based on 100 per cent of the population are probably less valuable than a large, biased, sample of mail-responsive persons for 1993. The statistical bias in the marketing database which may render the data of reduced academic value in certain applications (e.g. for estimating numbers of 5–10 year olds) is now a positive advantage – 5 to 10 year olds do not, after all, order by mail! In many marketing applications the population of interest is that fraction of the total census population which is likely to be mail-order responsive. If the database is biased against the poor, the homeless, and the extremely rich, then this might actually make it *more*, rather than less, valuable.

Another key aspect is whether or not all of the data are postcoded/zipcoded. In the UK, there is an almost universal use of postcodes with address data, the principal exception being the central government Office of Population Censuses and Surveys (OPCS)! Postcodes make it very easy to introduce a geographical dimension to address data and thus to add value to it (Raper *et al.* 1992). For example, a list of names and addresses input from the electoral roll has no value. Once postcoded, other information can be added to it using geographic file linkage (i.e. a geodemographic code) and it becomes a valuable commodity. However, what is perhaps a little surprising is that whilst postcodes can be converted into geographical coordinates most users do very little with the mapped information. Location is clearly important, but apart from drawing maps of such things as the distribution of clients, it is not really used. Maybe this reflects the GIS revolution of the late 1980s and the popular appeal of mapping systems. Indeed, most business applications for GIS seem to be largely mapping orientated. Dangermond (1993) lists what he terms several major areas of business: real estate, energy, electrical utilities, forest products, tourism and recreation, transportation, communications, publishing, and insurance. He writes:

> In just the last few years the application of GIS to business has begun to grow very rapidly. This seems to be because the technology is getting significantly easier to use, much less expensive, and because the business community now accepts the value of GIS in performing many kinds of necessary work.   (p. 9)

There is no doubt also that there is a growing number of PC-based marketing systems that allow various GIS operations to be performed on marketing data (for example, see Cresswell, Chapter 9, this volume). It is fairly easy to plot maps of clients or market penetration rates against various map backcloths: for example, a UK Ordnance Survey (OS) map base, or street-lines, or travel times, etc. Various site evaluations can be performed – for example, population or potential sales profiles within x km of a retail site – and, indeed, this can be a highly profitable and popular activity. Dangermond (1993) notes that this can be useful as a means of locating new business. However, it is also a very 'Noddy' spatial analysis approach to what is a complex activity and provides no real basis for multi-million dollar investments (e.g. see Birkin, Chapter 6, this volume). Superficial site evaluations and displays of customer and outlet locations may look picturesque, and spatial query may be helpful in explaining what is going on to senior company executives, but it often has no other value. Spatial decision-support using GIS often reduces to no more than just eye-balling map displays, or querying spatial databases by pointing a mouse. At the very least you need a 'what if' capability and ideally much more as well. A desktop mapping-come-GIS does not provide a cheap route to spatial decision-support although it can aid understanding and prepare the way for acquisition of better systems.

The problem here is that the rapidity of development of marketing databases and information systems has outstripped our ability to make good use of them. If a marketeer knows exactly what profile clients should have then there really is no problem. For example, you might want 100 000 names and addresses of golfers aged 40 to 55 with £30 000 plus income, as being the ideal profile for a given product. However, performance now depends on (a) you being right about this ideal profile, (b) a good performance of a very simple buying model that assumes people of this profile will respond at a much greater rate than other types of people, and (c) accuracy of the database variables used in the selection. In practice, much must depend on the identification of the target profile, but can you be that precise or that confident? What if you are wrong about your customer base, or don't know? It does not help to know that you could also have made selections based on at least 57 other variables as well! Finally, why are you not using the geographical coordinates as an additional targeting aid?

7.3.3 Emerging new marketing technologies

Clearly then data are being generated faster than they can be digested. Very large (over 50 gigabyte) marketing databases now exist and provide a plethora of information: some may be useful, some may be totally useless, but this division is context-dependent. There are details of personal and household characteristics, of consumer behaviour, of lifestyle variables, of wealth and possessions, and of behaviour; and much, much more. Yet what do we do with them? Most databases exist on hosts that can perform simple queries and print address labels, probably in a mainframe and batch environment. Yet the speed-up in computing and the revolution in computing economics make it feasible to apply dramatically more complex and intelligent analysis procedures than are currently practised. There are emerging new ways of mining databases for far more effective variable selections and overall database performance. The technological basis for marketing has started to change rapidly: and yet most marketeers still seem to be using old-fashioned and even obsolete 1960s technology, aimed at solving problems from the 1970s, clothed in 1980s GIS, and run on 1990s hardware. There is a danger that many are using simple, sloppy, 1960s systems because they yield understandable results.

Openshaw (1992) terms this the Catherine Cookson approach to marketing. At one time there was no other choice; simple methods, understandable by non-technically minded users, reigned supreme. However, the new technologies are inherently more complex and are increasingly incapable of yielding simple results. The problem is how to use them safely and painlessly despite their complexity. This can be done if effective use is made of graphical interfaces. Most car drivers have no good idea as to how their car works but this does not stop them either driving cars or trusting them. The equivalent systems in marketing also need to be user-proofed and rendered intrinsically safe. Once, this required extensive statistical and mathematical skills, but now it merely needs an automated process with built-in checking and validation mechanisms. The extra computing power can be used to power a fail-safe, almost fool-proof technology. Openshaw (1992) discusses this further.

One aspect of changes in marketing and marketing technologies is the greater emphasis being placed on statistical methods. However, this amounts to little more than a catching-up process, whereby the less sophisticated increase their levels of statistical competence. It is also a filtering-down process, as once 'advanced' methods become sufficiently affordable to be more widely applied. None of this is particularly relevant to marketing and there are problems with the statistical tools, often due to the non-ideal nature of the data. Moreover, it is difficult to build stable models of what are always low probability events (i.e. success in a marketing context typically has a response probability of less than 0.05, and often less than 0.001). Finally it is questionable whether statistical methods make the best or even just good use of the available data.

The current enthusiasm is for neural nets and perhaps soon bio-marketing: see Furness (1992), Barrow (1993), and Openshaw (1994b). The problems tend to be severe when a largely non-technical user community begins to combine very large and massively multivariate databases with advanced analysis and modelling tools that offer good performance at the expense of brittleness. The growing list of available methods (scoring methods, regression, discriminant analysis, logit models, cluster analysis, AID, CHAID, and neural nets) does nothing to help either. Moreover, even after all this apparent sophistication, there is still no obvious spatial-analysis component to the results.

In a modelling and response-prediction context Openshaw (1992) has argued that developing black-box automated analysis tools might be the only safe solution. That is to say, if there is uncertainty as to which methods to use, then try them all. His Marketing Machine (MM/1) concept offers a conceptual framework for this approach. It recognizes that the most important factor is 'safety' and this puts considerable emphasis on evaluating the results and checking for good, robust performance. This approach may not be very intellectually satisfying but would seem to be a pragmatic and extensible solution to an emerging problem. However, it is still not spatial in any way, even if it does provide a framework within which spatial analysis tools might one day be applied.

7.4 Spatial analysis in marketing

7.4.1 Where are the opportunities?

Merely because marketing data have implicit geographical referencing does not automatically mean that spatial analysis is needed. Discovering how to exploit the spatial element is not easy. Location may well be considered by geographers to be an *a priori* important variable in everything they do. The problem is that you cannot simply use (x, y) coordinates as merely another pair of predictor variables in a regression model. This has been attempted, but it will seldom be sensible in a marketing context. Instead, one obvious approach is to try and utilize the great strength of spatial data by focusing on methods of analysis that look for geographical patterns at regional and local scales. A common and recurrent aspect of geographical information of all kinds is that there are strong, often localized, patterns. These spatial heterogeneities or non-stationarities or localized extra-Poisson variation (or whatever else they might be called) can present horrendous problems to conventional statistical models (which assume global relationships without many or indeed any spatial singularities). Yet other methods, often quite simple ones, can be used to define these patterns. It is argued, therefore, that the most useful spatial-analysis techniques for marketeers will be procedures for describing spatial patterns. Table 7.1 lists some of these methods. Note that mapping raw data is not analysis and is not sufficient by itself. However, once the data are processed to separate the 'rough' from the 'smooth' components then mapping becomes much more useful. It is with spatial pattern recognition and modelling methods that the main opportunities for spatial analysis lie in a marketing context.

Table 7.1 Spatial analysis tools relevant to marketing

---

spatial classification
spatial cluster detectors
fuzzy spatial targeting
spatial relationship seekers and modellers
smart modelling
spatial pattern recognition
spatial data value adding

7.4.2 Spatial classification

The existing geodemographic systems are one example of spatial pattern seeking. Clusters of geographically contiguous census areas provide useful descriptions of multivariate census data. Space is being used here as a convenient metric for simplifying complex data. The 'old' geodemographic argument is applied so that it is assumed that areas with similar characteristics might well behave similarly in response to marketing stimulus or share similar needs. However, to make the most of the benefits of geodemographics the classifications need to handle rather than ignore both the varying spatial reliability of the data (because of small-number problems) and the varying precision of the variables themselves (see Openshaw 1992). The smaller the spatial units the greater these problems become.

7.4.3 Spatial cluster detectors

An obvious development of crude spatial classification technology is to focus more on the clustering inherent in client data or in response to direct mail campaigns. It is interesting that spatial epidemiological methods have seemingly not been used much, or at all, in marketing, despite a strong implicit recognition that positive response is a highly contagious process (i.e. neighbourhood effects). If responders cluster at some spatial scale, then location can be used as a filter on the selection process. Also, the locations of the clusters may themselves be of marketing relevance: if recurrent meta-patterns can be identified, then they have an obvious use in response prediction.

The underlying spatial analysis assumption is that socio-economic characteristics are not by themselves sufficient to describe or predict response, and that it is interaction of these variables with relative geographic location that is most critically important. These interactions are of two kinds: neighbourhood effects and location effects. The former is dealt with to some extent by spatial classification but might also be made more explicit. The assumption here is that responders cluster, more or less regardless of socio-economic characteristics. The relative location effects are probably either evidence of recurrent patterns or spatial proxies for missing variables – for example, place, culture, accessibility, locality, etc. Examples of relative location effects would include: proximity to London's orbital motorway ring, rural not urban, town centre, peri-urban, seaside location, market town rather than industrial town, small town rather than large town, New Town, high/low unemployment area, areas of recent population influx, etc. A systematic search might well be worthwhile, especially if it could cover a large number of marketing response data-sets.

The cluster detection tools of relevance here are those developed to detect localized anomalies in rare cancer databases: see Openshaw and Craft (1991); Besag and Newell (1991); and Cuzick and Edwards (1990). These technologies are of considerable potential value in marketing in the UK, especially once the Ordnance Survey's Address Point product provides accurate locational referencing of all addresses in the UK (see Longley and Martin, Chapter 2, this volume). However, even without it, considerable progress could be made and new dimensions added to direct-marketing targeting systems. This spatial epidemiological technology is particularly appealing because even the lowest marketing response rates will exceed rare cancer incidence by at least one order of magnitude. It is also apparent that the marketing industry could do themselves a great favour by mining their data-sets for clustering in response to their mail-shots, with a particular concern for persistent distributions, especially clusters, that are in predictable locations, or clusters which are common to different types of products. In general, they might well be able to become smarter by switching to more flexible technology. The artificial life methods described in Openshaw (1994a) can be used to pattern-hunt in marketing databases, with only a minuscule amount of pre-selection. Indeed this is already the subject of work in progress.

7.4.4 Fuzzy spatial targeting

The concept of a fuzzy geodemographic system is not new (Openshaw 1989a, 1989b), although none exists as yet. The basic idea is to exploit the spatial fuzziness in the standard geodemographic systems. There are two areas of fuzziness: first in the assignment of a census tract or enumeration district (ED) to a cluster and secondly in the assignment of a postal address to the census enumeration district (see Martin and Longley, Chapter 2, this volume).

In all multivariate classifications of spatial data it is unavoidable that some areas will be mis-assigned, and others may well be 'near' in a taxonomic sense to more than one cluster. Yet the current systems are based on all-or-nothing assignments, so no fuzziness exists. A fuzzy geodemographic system would allow the user to question which addresses are not actually in the target groups but are 'near' to them in the similarity space of the classification. This is important in a spatial context because the internal heterogeneity of census tracts is so great.

The second source of fuzziness concerns the assignment of addresses to census tracts. Currently, there is no 100 per cent accurate directory, but even where directories do exist, it might still be beneficial to consider fuzzy queries, such as how to identify addresses which are not actually in the target areas but which are within 250 metres of them. Obviously both sources of fuzziness can be combined to yield answers to queries such as how to find addresses within 500 m of target areas within clusters which are different from, but near to, the target ones.

The problem which then arises is that all of this requires much larger databases and much more complex user interfaces. The ideal system would automatically calibrate itself by identifying optimal levels of fuzziness for any particular marketing application. The benefit offered here is a rejuvenation of geodemographics by targeting areas previously excluded by 'all or nothing' approaches.

7.4.5 Spatial relationships seekers and modellers

The other value of location is as a proxy for other missing variables. For example, a high response from addresses near to a railway may well reflect either accessibility difficulties due to the barrier effect of the track or be a proxy for some socio-economic, community or cultural effects. It is important to pick up local spatial associations that seem to matter. Whilst the focus on individuals rather than areas in targeting may have been very beneficial, the opportunities for further segmenting individuals on the basis of their spatial relationships with surrogate map features might also be useful.

The problems here are purely methodological. If you wish to use spatial relationships as surrogates then you need an exploratory analysis technology to help you do it. It is interesting to note, therefore, that the Geographical Correlates Exploration Machine (GCEM) of Openshaw *et al.* (1990) started out as a market research tool. Maybe it is time to re-apply it.

A different type of spatial relationship might be found in consumer responses to regionalized variables. Spatial trend surface regression methods have been used in research for at least a decade. It is possible that these methods may also be useful at more macro-geographic scales as a means of modelling spatial response. A good review of relevant methods is provided in Haining (1990). Spatial regression methods should also be used in preference to conventional regression methods whenever zonal data (e.g. postcode sectors) are being analysed, especially if there is concern about significance testing or interpreting regression parameters by reference to their standard errors.

7.4.6 Smart modelling

The real leading edge today lies in the development of smart modelling and analysis systems. Market research is an industry that quite often cannot see the forest for the trees. Users need to find or discover insightful, interesting, and commercially valuable patterns or relationships without necessarily being able to define in advance what they are looking for. They know what they want, it is just that they do not know how to get it in anything like an optimal manner. The required paradigm is quite straightforward. You input all the available data, you provide a sample of successes and failures, and you hope the system will do the rest. Amongst the input data will be geographical variables: the system, not the user, determines whether space matters and, if so, how best to incorporate it. You could use a neural net, but quite often better results will be obtained by using a much simpler memory-based reasoning paradigm, a form of case-based machine learning. Basically, you either seek to classify cases against a library of good and bad profiles noting k^{th} nearest matches, or you build up a library of good performers. There are various ways of achieving both objectives. The point to note here is that the results can be extremely good in situations where no other model would work at all. You are using the knowledge that is acquired as a result of the marketing to improve the performance of subsequent marketing efforts.

An interesting variant of this paradigm is to allow a genetic algorithm (GA) to create good customer profiles from the library database. In essence the GA breeds query operators which would result in clusters of customers with the highest response rates. This might well be considered to be an artificial life approach. How you explain the results to marketing end-users is left to the reader's imagination!

Another important aspect concerns the need for feedback. The ideal systems are adaptive and dynamic. They are not based on any understanding of detailed process, because this is too difficult and we understand too little to do this at present. Ideal systems sense their environment and they respond to it without being told how to do it. The Alife (artificial life) clients are 'alive'; they are living symbolic creatures. They will adapt to change. They are ideally placed for real-time systems or for marketing operations that generate feedback. In a phased mail-shot you start from a position of little or no information, you gain response information and the targeting should improve if the generation of subsequent mail-shots is sensitive to success and failure, as well as exogenous changes. The technology exists, although it is just not used much at present (Openshaw 1994b).

7.4.7 Spatial pattern recognition

A final area of potential spatial-analysis relevance concerns what might be termed spatial pattern recognition. There is a risk in marketing that major spatial and aspatial regularities are being overlooked. Perhaps the available analysis tools are too sophisticated and are attempting to be too clever and too smart. By focusing on the minutiae we may be failing to see the whole. Instead of trying to predict response for item x or campaign k, maybe we should be considering the overall spatial pattern of response. Are we simply seeing different versions of the same generic pattern or different spatial patterns? If the latter, is there a library of different spatial pattern types that might exist? Maybe there is a macro-scale geographic pattern to most responses, and if there is and if it can be parameterized, then there is a basis here for an entirely new approach – that is to say, there is a macro spatial pattern framework within which to fit the detail. It would certainly be interesting to investigate this further.

7.4.8 Spatial data value adding

The final use of spatial analysis in marketing is as a means of enhancing databases by geo-processing and linking in other spatial data (see Table 7.2). Geodemographics is one example. More importantly there is now a tremendous wealth of digital map data – in the UK case, from the Ordnance Survey, the Automobile Association, and Bartholomews. Much of the information is not directly relevant but some is. In particular, the various feature overlays in the digital map databases can be used to tag postcode and census data with contextual information such as distance to various linear features (coastline, certain land uses, rivers, woods by type, golf courses, urban boundaries), or location within polygons of interest (e.g. built-up areas) or distance to polygon of interest (e.g. parks).

Table 7.2 Adding value to marketing data by geoprocessing

distance to selected digital map features
locations within selected digital map polygons
non-standard aggregation
place name processing
house name processing

In addition, there are other ways of adding value by geo-processing, principally by the analysis of addresses for place-name and house-name content. There is almost certainly an extremely high association between social status, wealth, income and postal address and, when present, house names. The associations are not straightforward, nor are they linear: but they are undoubtedly strong. Modern GIS and computer technologies provide the tools for investigating further these spatial associations – if only to add new layers in the spatial query process and new predictor variables in the response prediction models. Indeed, as the latter become cleverer and less dependent on human beings to provide all the intelligence in the form of specification and re-coding of variables, so it becomes possible to make use of the wider range of contextual spatial information that is available or could easily be generated (see also Openshaw 1993b).

7.5 Conclusions

This chapter has attempted to stimulate interest in applied spatial analysis relevant to marketing among both geographers and marketeers. Similar comments might also be made about other areas of potential spatial-analysis application. However, marketing has the distinction of being the best commercial prospect for exploiting spatial-analysis skills and tools. A number of generic spatial-analysis approaches have been defined as being of potential and real marketing relevance. Now is the time to try them out. Do they work? Can they be made to work? What further developments are needed? What further research is needed? How can we best respond to the needs of the end users? These are all challenges for quantitative geographers. However, the process does not stop at the research stage. Promising methods need to be converted into useful products so that they can be sold or consultancy services can be created around them. From a profit-generation and value-added perspective this latter path has the greatest potential, but for it to work it requires a combination of the 'right' environment with persons who *want* it to be successful. This is still a rare occurrence, suggesting that a major culture change is required.

It should be appreciated that geographers have no proprietorial rights in this area: they are not that unique in what analysis skills they offer. Whilst it might take five years or even a lifetime to convert a geographer into a 'good' computer scientist, it may only take a few months for a computer scientist to learn or pick up the bits of geography that matter. To this extent, time is running out and, as spatial analysis tools are eventually packaged for popular application, so the opportunity could slip by. This may, of course, be no bad thing but equally it might also be a significant waste of new research opportunities, a turning away from new areas of scientific investigation and scholarship, and a neglect of the possible commercialization gains from technology transfer.

Part Three

Proprietary and Customized GIS for Business and Service Planning

Although geodemographic analysis has hitherto been regarded as the mainstay of GIS in business and service planning, there has been a gradual realization that such indicators can provide only a partial picture of geographical markets. While they may be reasonably adept at describing 'what is' patterns, they are inadequate at answering 'what if' questions concerning future business tactics and strategies. There has thus been a call for greater spatial analysis capabilities within GIS in order to respond to these tactical and strategic issues. In addition, many of the early uses of GIS in business and service planning have been restricted to routine visualization of problems. While a useful starting point, there is currently some disillusionment and disappointment that the potential and promise of GIS in geographical analysis has not been fulfilled. A point repeatedly made by practitioners is that after the impact of pretty pictures has worn off, GIS seem barren: business solutions require *analysis*, and not just visualization of problems.

How has the GIS industry responded? The industry is huge and is developing and diversifying with extreme rapidity: hence there is no simple or single answer to this question. The three contributions to Part Three are written by geographers working in rather different sectors of the GIS industry, and provide a range of perspectives. David Maguire is the UK Technical Director of a large GIS vendor, and in Chapter 8 he describes and evaluates a range of relevant analytical operations which may be carried out using proprietary GIS. He also describes how such systems now offer linkage or 'coupling' to the sorts of mathematical and statistical packages with which market researchers and analysts may already have some familiarity. He thus provides a well-rounded discussion of proprietary GIS, and the increasingly flexible way in which they are amenable to different data structures and analytical problems. He is careful to point out that academic geographers have a tendency to focus upon problems of apparent intrinsic academic merit and complexity, while paying scant regard to less glamorous user needs in the 'real world'. In so doing, he echoes some of the sentiments of Stan Openshaw in Chapter 7.

The second contribution is by Paul Cresswell, who is a director of SPA Marketing Systems. He begins by describing how the plummeting real cost of computer processing power led to the transformation of a mainframe-based bureau service into a consultancy which supplies PC-based systems to a range of commercial clients. The market niche which is described here is therefore the supply of customized GIS products to particular clients, where customization may frequently use elements of proprietary software as the basic building blocks. Case studies are used to describe GIS solutions to problems of site location, acquisitions and mergers, merchandizing, and performance analysis. The range of analyses that he describes include geodemographic applications of the sorts developed by the contributors to Part Two, and analyses based upon postal geography as described in Part One. The merits of customized solutions are set out in terms of cost, ease of use, flexibility and functionality, and (like Maguire in Chapter 8) he is careful not to dismiss the continued importance of some mapping operations.

The final contribution to Part Three, by Graham Clarke and Martin Clarke, adopts the position that many clients require no less than a wholly customized solution to business or service planning problems, and that proprietary software is often too inflexible to respond to these demands. The GIS solution that they advocate is thus for consultants to become immersed in the specific problems of their clients. Resultant GIS solutions may frequently require the writing of customized software systems from scratch, or at the very least of developing very close coupling to specialist mathematical and statistical routines. They also attempt to quantify the immense financial benefits that can accrue when businesses adopt spatial models within a GIS framework.

The particular GIS strategy adopted by any organization is likely to depend upon a number of factors. Each of these solutions presumes a different mix of hardware, software (e.g. system acquisition) and personnel investments, and different strategies will be more or less appropriate to operational versus tactical/strategic requirements. Each also has relative costs and benefits, although the benefits of strategic use may be less readily quantifiable, and hence accountable, than the benefits for routine marketing functions. All of this is but a small beginning, with business and service applications likely to be one of the fastest growth sectors in the GIS industry over the next decade. The message is clear: GIS has the potential to fulfil an integrative role throughout business and service corporate strategy. There is already evidence that GIS applications are extending from the operational functions of businesses and services towards the tactical and strategic levels, and that many companies require GIS inputs to strategic investment decisions at boardroom level: this is an argument developed in greater length by Birkin *et al.* (1995). There is also evidence that some firms view customization and consultancy as a means of 'future proofing' through commitment to regular updating of GIS capabilities.

Each of these contributions is informed by geographical training, in that each is written by geographers. This continuing relationship between the research base and applications-led research provides the theme to Part Four.

8

Implementing spatial analysis and GIS applications for business and service planning

David Maguire

8.1 Introduction

Over the past few years there has been a remarkable increase in interest in GIS. Many of the earliest users were in universities, government departments and environmental agencies. Activity in these traditional core areas is now being supplemented by vigorous growth in several emerging markets, the most important one being business and service planning. For many of these new users, the GIS focus to date has been basic mapping and asset management. Other, more advanced, users are modelling data held in integrated databases. This modelling activity is frequently referred to as spatial analysis.

This chapter is primarily concerned with the implementation of spatial analysis functionality in commercial GIS software systems and its implications for business and service planning. It looks at various approaches to implementing spatial analysis functionality and the application of spatial analysis in two of the most widely used general-purpose proprietary GIS software systems. The first, Environmental Systems Research Institute Inc's (ESRI) ARC/INFO, is a hybrid file–database vector and raster system and this chapter is written by ESRI (UK)'s Technical Director. The second, ERDAS's IMAGINE, is dominantly a file-based raster system. Other systems, such as TransCAD's Calliper, the IDRISI Project's IDRISI, Intergraph's MGE, GENAMAP, and the public domain GRASS (originally developed by the US Corps of Engineers) have, in some cases, reasonably sophisticated spatial analysis capabilities, but they are not as widely used or well known as ARC/INFO and ERDAS.

In the first section the development of spatial analysis is briefly reviewed. Next there is discussion of spatial analysis, and business and service planning. This is followed by a description of the current GIS software data models. After this is a section dealing with the spatial analysis capabilities in ARC/INFO and IMAGINE. In the following section different approaches to linking

GIS and spatial analysis software are considered. Finally, there are some brief conclusions.

8.2 What is spatial analysis?

'Spatial analysis' is one of those terms that are so widely used in so many different contexts that it is difficult to define succinctly. Goodchild (1988, p. 68) offers a good general definition of spatial analysis as 'that set of analytical methods which require access both to the attributes of the objects under study and to their locational information'. Openshaw (1991b, p. 18; also, Chapter 7, this volume) suggests that what geographers refer to as 'spatial analysis', statisticians call 'spatial statistics'. Anselin (1989) and Goodchild *et al.* (1992) prefer to use the term 'spatial data analysis' although there seems to be no substantial difference.

Openshaw (1991a, 1991b) argues that the generally accepted origins of spatial analysis lie in the development of quantitative and statistical geography in the 1950s. At this time, the focus in academia was on nomothetic (general) approaches to scientific endeavour. Kubo (1991), however, suggests that Japanese geographers were pioneering the use of spatial analysis in the 1930s. In the 1960s, the quantitative revolution in geography focused interest on point and quadrant analysis, and the application of standard linear multivariate analysis methods to zonal data (see Berry and Marble 1968 for examples of contemporary work). The late 1960s and early 1970s saw a decade of great interest and widespread use of spatial analytical methods by geographers and, to a lesser extent, other environmental and social scientists. During this period, the earlier work on statistical methods was widened to include mathematical model-building with an emphasis on theoretical development rather than on application. In sharp contrast, during the 1980s spatial analysis was largely forgotten by geographers as Marxist and humanistic interests came to the fore. Fortunately, however, there was still considerable interest outside geography, for example by spatial statisticians: see, for example, Ripley (1981), Diggle (1983) and Upton and Fingleton (1985). From the low point of the mid-1980s interest in spatial analysis has increased remarkably in the 1990s, largely on the back of the great upsurge in interest in GIS.

Interest in spatial analysis in the 1990s can be measured in many ways: a few examples will serve to illustrate this point. The US National Science Foundation 'Request for Proposals' for establishment of a National Center for Geographic Information and Analysis (NCGIA) lists five areas to be addressed, the first of which is to 'improve methods of spatial analysis and advances in spatial statistics' (Abler 1987). Goal one of the Dutch equivalent, NEXPRI (NEXPRI 1989), is 'research into the development of GIS and spatial analysis'. The National Museum of Cartography set up in Japan in 1988 will, as one of its five major functions, act as a research centre on geographical information and spatial analysis (Kubo 1991). Spatial analysis has also featured high on the UK ESRC Regional Research Laboratory

Initiative agenda (see Clarke *et al.*, Chapter 12, this volume). Finally, 'spatial analysis' gets 41 mentions in the index of the 'Big Book' (Maguire *et al.* 1991), placing it fifth behind 'raster', 'remote sensing', 'United States Geological Survey' and 'vector'.

All this activity has even caused some commentators to proclaim that spatial analysis has been reinvented by GIS (Openshaw 1991b). The reasons why GIS has contributed to the rediscovery of spatial analysis are many and varied; some of the more significant from the standpoint of proprietary GIS usage by business are as follows (see also Openshaw, Chapter 7, this volume).

1. The last few years have witnessed huge improvements in the price : performance ratio of computer hardware, such that it is now possible to package very high-performance processors in the form of desktop personal computers. This has substantially removed the processing bottleneck restricting spatial analysis which was evident even just a few years ago.
2. Recently, there has also been a massive proliferation of data about the environment and society brought about largely by advances in data collection methods and technology (e.g. satellite remote sensing, Global Positioning Systems and Electronic Point of Sales in retailing). This has required a major shift in spatial analysis from ideas based on a situation which is data poor to one which is data rich.
3. GIS also provides spatial analysts two sets of very important tools that allow excellent data management and visualization. It is now inconceivable for spatial analysts to work without access to good tools for data management (e.g. some type of Database Management System – DBMS – software) and visualization (e.g. graphical and cartographic drawing software).

These technological developments have led both to a great general increase in using geography as an organizing framework and to the commercialization of GIS. In spite of all the hype and activity, there is, nevertheless, a general feeling that most GIS systems are even today still used for data capture and database automation. Openshaw (1991b, p. 19) even goes as far as suggesting that 'most GIS offer little or no spatial analysis technology. A typical GIS will contain over 1000 commands but none will be concerned with what would traditionally be regarded as "spatial analysis".'

More generally, Fotheringham and Rogerson (1993, 1994) consider eight impediments that arise in spatial analysis and some of the ways that these have been tackled using GIS. The impediments they identify are summarized here because of their relevance to implementing spatial analysis for business and service planning in commercial GIS.

8.2.1 Modifiable Area Unit Problem (MAUP)

Simply stated, MAUP arises because different types and levels of aggregation can produce wholly different representations of geographical phenomena. GIS have the capability to reorganize data rapidly to examine the sensitivity of different zoning systems and thus can assist spatial statisticians. For example, when undertaking geodemographic analysis data are frequently available at several levels of aggregation (see Waters, Chapter 3, this volume, for an analysis of the scales at which European census and other public-sector data are made available). Because of MAUP it is likely that any results from integrating, say, census data with customer information will be influenced by the level of aggregation selected (see Martin, Chapter 4, this volume).

8.2.2 Boundary problems

These occur because geographical study areas are usually bounded in ways that do not correspond with the effects of spatial processes. Thus features just outside the study area, although important factors affecting the processes under study, may not be included. Using GIS it is possible to visualize data for the purpose of identifying boundary problems and to examine the impact of different boundaries by re-running analyses with different features excluded. Boundary problems might come into play in undertaking location–allocation analysis on, say, some schools. This might involve locating a series of schools in a new urban area and then allocation of pupils to the schools. Unless a study area of sufficient size is chosen, boundary effects may influence the results.

8.2.3 Spatial interpolation

There are many different methods for interpolating and extrapolating geographical phenomena. The selection of methods depends on the type of data, the degree of accuracy desired and the amount of computational effort afforded. Clearly, the visualization, data management and computational capabilities of GIS will assist in comparing and selecting appropriate methods. Spatial interpolation is a very important spatial analytical technique for business and service planning. Such techniques might be used, for example, where an analyst wishes to create a continuous cost surface from point samples showing household expenditure on certain goods and services. The method of interpolating the surface from the point data will certainly influence the final result.

8.2.4 Spatial sampling

Over the years many spatial sampling strategies have been developed. Whilst the appropriateness and accuracy of the main broad types is well known, the nuances of unusual strategies are not. GIS could be used to evaluate the ability of different strategies to capture the variation of different phenomena effectively. All data used in business and service planning is a sample of the total possible values which are available. For example, customer databases are only – usually biased – samples of all possible consumers. However, Openshaw (Chapter 7, this volume) and Birkin (Chapter 6, this volume) argue that the bias inherent in business databases does not necessarily create severe impediments to market analysis.

8.2.5 Spatial autocorrelation

It has been known for some time that spatial processes exhibit patterns which invalidate some of the basic assumptions of aspatial statistics. Spatial autocorrelation provides useful measures and descriptions of the pattern of spatial processes. GIS can be used to manage data in a format suitable for calculating spatial autocorrelation and also to provide convenient methods of visualizing results. In a general sense, spatial autocorrelation can be described as a realization of the fact that everything is related to everything else, but near things are more inter-related than distant things. It is, for example, much more likely, other things being equal, that two households which are next to each other will have similar disposable incomes than two households which are far apart (see Openshaw's discussion, Chapter 7, this volume, of spatial clustering). Unfortunately, many of the simple classical statistical models which are routinely used in spatial analysis (e.g. regression and correlation) assume that there is an equal probability that houses close together or far apart will have the same disposable income. This renders the simplistic application of such models open to misinterpretation.

8.2.6 Goodness-of-fit

Many statistical methods use the goodness-of-fit (a measure of how well a model matches some data) to calibrate the methods and infer processes. GIS offer the opportunity to compare different goodness-of-fit techniques by mapping the results and also to control the spatial resolution of matching. In business and service planning it is a frequent requirement to model the accessibility of stores to customers using spatial-interaction modelling techniques. There are several types of models which can be applied and several methods of evaluating the results using goodness-of-fit statistics. The choice of model specification and evaluation statistic may have a significant influence on the conclusions which can be drawn.

8.2.7 Context-dependent results

In spatial modelling it is often the case that parameter estimates vary significantly over space when they are calibrated separately by location. By embedding such models in GIS it should prove easier to accommodate this problem using the computational and data-display capabilities of GIS. Again this problem can be illustrated using an example based on buffering operations. It might be assumed in spatial buffering operations, such as when examining the pattern of consumers visiting a hardware store, that there will be an equal fall off of customers with increasing distance in all directions. If, however, the cost of travel is not symmetrical in all directions this needs to be reflected in a modified distance decay function (see Birkin, Chapter 6, this volume, for a fuller discussion of this problem). The incorporation of multiple data sets and the ability to recalculate operations rapidly will greatly facilitate investigations into this phenomenon, which will in turn lead to more accurate results from such operations.

8.2.8 Aggregate versus disaggregate, and simple versus complex models

In the early stages of spatial modelling, analysts normally have to decide whether to use aggregate or disaggregate data, and whether to go for a simple or complex modelling strategy. With the advent of GIS that have powerful data storage, management, and display capabilities, it is now possible to evaluate more closely which is the most appropriate strategy by comparing the results of several levels of aggregation and complexity. For example, models of migration frequently work at the aggregate level, and utilize sum total flows between census areas. An equally valid, and in some cases more useful, approach is to look at the migration of individuals (after all, decisions about migration are seldom made by communities). A system which could undertake both levels of analysis and perhaps even integrate them together would thus be of enormous benefit to migration analysts.

Clearly, interest in spatial analysis as a research topic has been enormous in recent years, with it featuring high on the research agenda of all major national initiatives. Academics and researchers have been quick to extol the virtues of spatial analysis and deride the use of GIS for inventory applications. It is interesting, therefore, to examine the impact of spatial analysis on the developers of GIS technology, and on the ultimate arbiters of the value of spatial analysis, the users of GIS.

8.3 Spatial analysis and business and service planning

The focus on data capture, database automation and basic inventory operations in business and service applications of GIS is not altogether surprising, given its relative youthfulness as an application area (see Sherwood, Chapter 11, this volume, for results of a survey of 'business geographics' users). Over time it might be expected that business and service planning applications will become more analysis-, modelling- and management-oriented. Such trends are commonplace, as the work of Crain and MacDonald (1984) and Maguire

(1991) shows. This work suggests that the initial reason for establishing most information systems and, therefore, the main activity in the initial development phase (Phase 1, Figure 8.1) is assembling, organizing and undertaking an inventory of features of interest (e.g. customers, transport networks, population or schools).

The second phase in the evolution of information systems arises from the desire to undertake more complex analytical operations. Frequently these require access to data from several disparate sources and the use of statistical and spatial-analytical techniques to integrate the data. For example, answering even an apparently simple question such as 'which areas have the highest percentage of my customers' will require data on customers (typically postcoded or zipcoded point data) to be integrated with geodemographic data (typically area-based population counts derived from a national population census). The process of integration may involve: (1) point-in-polygon analysis to determine which customers fall within which census area; (2) calculation of the sum total of customers in each area; (3) standardization of the customer counts by dividing by the total population; and (4) display as a choropleth (shaded area) map and a report showing statistical data (see Martin and Longley, Chapter 2, this volume).

The third and most developed phase sees the evolution of an information system from a transaction-processing to a decision-support system capable of sophisticated analysis and modelling operations. During this phase there is considerable emphasis on spatial analysis and modelling. Typical applications in a business and service planning context include: determining which hospital or school is to close, given information about patients/pupils, the location of other hospitals/schools, an assessment of likely demand derived from population information, etc.; estimating the optimum route and warehouse location for a trucking company on the basis of information about the volume of goods to be transported, the location of customers and the characteristics of the transportation network; and evaluation of alternative land-use patterns for a new city development, given access to information about the geology and soils, existing land use, transportation routes and planning consents.

8.4 Current GIS data models

It is possible to classify the current crop of commercial GIS software systems in several ways. A convenient classification, as far as software developers are concerned, is based on the data management architecture (Maguire and Dangermond 1991). Essentially, data may be stored either in simply structured 'flat files' or in the proprietary format of commercial database management systems (DBMS). This leads to a three-fold classification: file processing, hybrid and extended DBMS (Figure 8.2: see also Martin, Chapter 4, this volume). This has direct relevance to the way in which spatial analysis can be implemented in GIS as we shall see later.

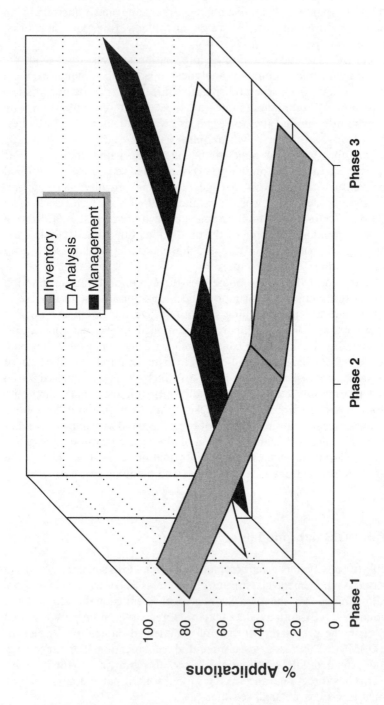

Fig. 8.1 The development of business and service planning GIS applications

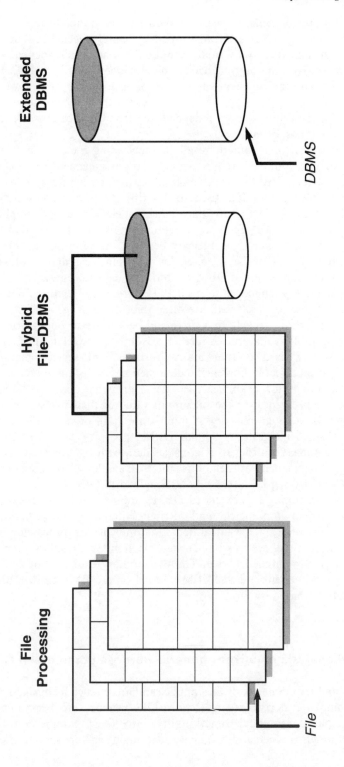

Fig. 8.2 GIS design models

In the simplest, file processing design, all data are stored in flat files. During GIS processing the data files are transformed by a user-selected function to create a new flat data file. Many small, simple, PC- and raster-oriented GIS software systems favour this approach; examples include IDRISI, ERDAS and IMAGINE. The data management capabilities of such systems are, however, limited. Users of such systems are not able to benefit from the advantages of DBMS, namely, multi-user access, independence from application software, security, and validation of updates.

An alternative design, favoured by all of the leading medium and large proprietary systems, utilizes the very good data-management capabilities offered by commercial DBMS for attribute data. In such hybrid systems the geometrical and topological data used to describe geography are stored in proprietary flat files, while the attribute data associated with geographical objects are stored in a DBMS. A key requirement is that the GIS should automatically maintain referential integrity between the geometrical, topological and attribute data. Flat files are used for the geometrical and topological data because of the limitations of DBMS in terms of data storage, access times and geographical query languages (Newell 1993). Presently, relational DBMS are preferred (in which case the GIS are often referred to as 'geo-relational'), but in the longer term it is likely that there will be a move to object-oriented DBMS as the technology matures. A further reason why the relational DBMS approach is widely used is that many business and service planning organizations have already implemented sometimes substantial corporate databases containing, for example, customers, property and assets. Such organizations are keen to lever the maximum value of their investment and analyse all their data geographically. A multi-store retail organization might, for example, wish to integrate its tabular DBMS property database with its GIS customer database in order to relate sales data with the volume of space allocated to each department at each of its stores. Examples of so-called hybrid GIS include ARC/INFO, GENAMAP and Smallworld GIS.

The third design employs an extended DBMS architecture. In such systems all data (geometrical, topological and attribute) are stored together in a DBMS. Although offering an integrated environment for processing and transaction management, systems built using this architecture suffer from the problems of DBMSs outlined above. The two primary examples of systems employing this architecture are SYSTEM9, based on the EMPRESS DBMS, and GEOVISION, based on Oracle.

8.5 Spatial analysis in current general-purpose commercial GIS

As the needs and requirements of GIS application areas (such as business and service planning) develop, more and more GIS vendors are beginning to incorporate spatial-analytical functionality into their commercial GIS products. Although the demand for such functionality is currently restricted to more experienced and sophisticated users, as others reach this stage so demand is expected to rise.

The developers of general-purpose commercial GIS are faced with a difficult decision when implementing spatial analysis functionality. On the one hand there is a need to develop generic functionality that will service the needs of a large congregation of surprisingly diverse users. On the other hand there is a need to create specific-purpose models and tests for tightly defined application areas. The following descriptions of spatial analysis functionality in two of the market-leading general-purpose GIS software systems show that it is possible to identify and develop a pool of generic techniques for many users.

8.5.1 Current spatial analysis functionality in ARC/INFO

ARC/INFO, from ESRI, is an example of a system employing the hybrid GIS design. It is one of the most popular GIS software systems. ARC/INFO was first released in 1982 and has been extended and substantially redeveloped on several occasions. Although best known as a vector-based system, it also has extensive complementary raster-processing capabilities. ARC/INFO runs on all popular work-stations running the UNIX and VMS operating systems.

Prior to the 6.0 release in 1991, spatial analysis in ARC/INFO was limited to basic data manipulation and simple descriptive statistics on spatial and attribute data. It would not be fair, however, to say that ARC/INFO had no utility for projects involving spatial analysis. Even at release 5.0.1, ARC/INFO possessed numerous tools for capturing, organizing, manipulating and displaying spatial data. The overlay, buffering, aggregation, classification (e.g. choropleth mapping), descriptive statistics, and graphic and cartographic display tools have all been widely used to assist in spatial analysis (Table 8.1). These remain extremely valuable tools in the spatial analyst's tool box.

In version 6.0 there were many enhancements to ARC/INFO in the area of spatial analysis. These tools were primarily packaged as part of the TIN and GRID extensions to the ARC/INFO Core. To date, the main spatial analysis release of ARC/INFO has been 6.1. In this release there were many enhancements to Core, GRID and TIN. Most notable were the updates to Network, which included spatial interaction modelling and a solution to the travelling salesman problem. Both network and crow-fly distances can be used in spatial interaction modelling. ESRI has now (January 1995) produced the 7.0.2 release and spatial analysis again features prominently. More functionality is planned in Network (location–allocation modelling) and, especially, GRID (multivariate classification, cluster analysis, shape analysis, dispersion modelling and others).

Plates 8.1 and 8.2 illustrate the uses of some of this functionality in the domain of business and service planning. Plate 8.1 shows a population surface for the county of Leicestershire, in the United Kingdom, interpolated from ward counts using ARC/INFO. The area-based data were first converted into a

Table 8.1 Some of the spatial analysis functionality in ARC/INFO

Release	Module	Feature
5.0.1	Core	Descriptive statistics
	Core	Data manipulation
	Core	Data visualization
	Core	Classification (e.g. chloropleth mapping)
	TIN	Descriptive statistics and interpolation
6.0.1	GRID	8 Focal commands
	GRID	12 Zonal commands
	GRID	> 20 Global commands
	GRID	Distance commands
	TIN	More interpolation
6.1	GRID	Regression and correlation
	GRID	Autocorrelation (Geary and Moran)
	GRID	Hydrological modelling tools
	GRID	Surface interpolation
	Network	Spatial Interaction Modelling
	Network	Routing (Travelling Salesman)
7.0	GRID	Multivariate classification (e.g. clustering)
	GRID	Shape analysis
	GRID	Dispersion modelling
	Network	Location–allocation modelling

Focal commands operate on individual cells, zonal commands operate on pixels with the same value, global commands operate on whole grids. Core is the base product, TIN, GRID and Network are integrated extensions licensed separately.

grid and then interpolated using a smoothing technique (cf. the 'customized' technique of Martin, Chapter 4, this volume, illustrated in Plates 4.1 and 4.2). The resulting grid was then contoured and colour-coded to highlight variations (top left). At larger scales, ward boundaries and place names are added to provide a spatial reference (top right). Contours can also be draped on the original areal data for comparative purposes (bottom). Plate 8.2 shows the results of some location–allocation modelling carried out for the city of Redlands, California. In this model, students have been allocated to existing schools on the basis of the travel time across the existing road network. Further analyses could be undertaken to examine the impact of opening or closing a school, or the impact of demographic changes on the demand for schools.

8.5.2 Current spatial analysis functionality in IMAGINE

IMAGINE, from ERDAS Inc., is a popular example of a GIS software system based on the file processing design. ERDAS first created a raster-based GIS system, called ERDAS, in 1979 (see Maguire 1992 for an extended discussion of this system). The IMAGINE system is an outgrowth of this earlier system, although it shares many of the same basic principles. IMAGINE remains a

Table 8.2 Some of the spatial analysis functionality in IMAGINE.

Image interpretation
    multivariate classification
        Fast Fourier Transformation (FFT)
        supervised and unsupervised classification
    image clustering
    convolution filtering
    PCA (Principal Components Analysis)
Image rectification
Image projection
Spatial modelling (over 150 operators and functions)
    e.g., arithmetic, bitwise, Boolean, colour, conditional, distance, exponential,
    focal, global, matrix, relational, statistical, surface and trigonometric operations
Terrain analysis
    surface interpolation and contouring

---

predominantly raster-oriented system focusing on images and image analysis. The latest, 8.1, release of the software does, however, have considerable raster-modelling and vector-processing capabilities. ERDAS IMAGINE runs on all popular UNIX work-stations.

The spatial-analytical functionality of IMAGINE may be grouped into the following areas: interpretation, rectification, projection, spatial modelling and terrain analysis (see Table 8.2). The main capability for conventional GIS analysis is provided by the Model Maker software (part of IMAGINE Spatial Modeler). Model Maker is a graphical editor for creating GIS and image-processing models by using a palette of object-based tools to place icons representing modelling operations on a blank page. Once created, the models can be run against the data to create new data files and associated statistical output. Like ARC/INFO, IMAGINE has a macro-programming language to assist in model development.

Figure 8.3 is an example of a sensitivity analysis model created using ERDAS IMAGINE. The purpose of the model is to identify the most environmentally sensitive regions in a study area so that they can be protected from development. There are three primary input rasters to the model (*n1_slope* – a raster showing the terrain classified into slope categories; *n2_floodplain* – a raster showing the flood plains; *n3_landcover* – a land cover classification derived from a satellite image). These data sources are all for the same area and have been co-registered. The data sources are combined using a conditional function to create a raster called *n4_sensitivity*. This composite raster has regions which have steep slopes (above 25 per cent), are on the flood plain, and have riparian or wetland land cover. A SPOT satellite image (*n8_summary*) is then enhanced to increase the sharpness and contrast by passing a kernel filter over it. The resulting image (*n7_spot_summary*) is then combined with the sensitivity raster (*n4_sensitivity*) to enhance the spatial resolution of the final raster. An input integer data source (*n14_integer*) is also added to scale the data values in the final raster (*n5_sensitivity_spot*) into a

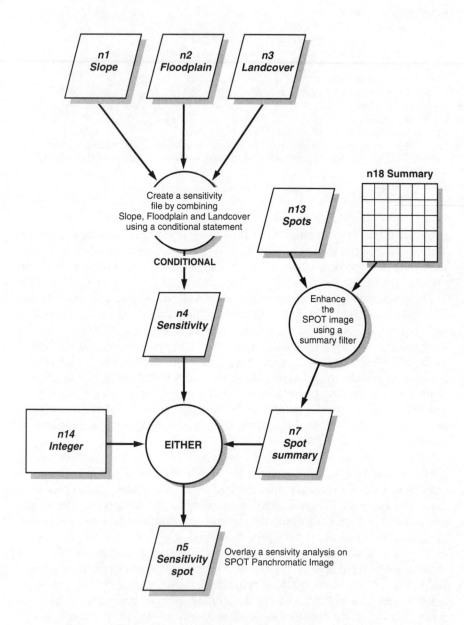

Fig. 8.3 A sensitivity analysis model created using ERDAS IMAGINE

meaningful range. This is a relatively simple model; much more complex models with a larger number of data sources and functional operations can be envisaged in a typical business and service planning application.

IMAGINE clearly has strengths in the areas of image processing and analysis, and environmental modelling. There are, for example, several alternative methods of image classification and multivariate clustering. This is not surprising given ERDAS's history in image processing and environmental analysis. In comparison to ARC/INFO it may be more limited in its business and service planning capabilities. Nevertheless, it does have a range of basic and some advanced spatial analysis capabilities. Business and service planning users might employ IMAGINE in situations where satellite imagery or aerial photography is required as a data source, either to provide background mapping or as a basis of land-use classification. IMAGINE also has utility for combining environmental and socio-economic data-sets for land-use evaluation and site suitability analysis.

8.6 Linking GIS and spatial analysis software

There is considerable interest in integrating GIS with other specialist systems to meet the requirements of advanced applications (Abel *et al.* 1994). The motivations for such work are that many of the current GIS have weaknesses for some advanced applications, and that combining GIS and other technologies will create new, richer approaches to problem solving.

There are many ways in which it is possible to link GIS and spatial analysis software, each of which has advantages and disadvantages depending on the particular application (Nyerges 1992). The basic factors which determine the approach adopted are:

1. Whether the systems being linked are extendable and sufficiently open to support data input/output.
2. Whether the GIS or the spatial analysis software is to be the dominant system.
3. Whether the final system must be fully integrated or whether simple data translation is adequate.
4. Whether information exchange between the GIS and the spatial analysis system should be dynamic or batch oriented.
5. The degree of effort the integrator (programmer) is prepared to expend to integrate the GIS and spatial analysis functionality.

8.6.1 Methods of interfacing GIS and spatial analysis software

Specific data translators

Many GIS software systems come with specific data translators as standard; ARC/INFO supports over twenty and IMAGINE over ten. Most of these are bi-directional and are designed to translate locational and, where appropriate, associated attribute data to or from the GIS's internal binary format. These translators work very well if both the GIS and the spatial analysis system support a common format and if only a batch link is required.

ASCII file read/write

Several of the GIS software systems with integral programming languages support reading and writing from ASCII files: ARC/INFO's AML and IMAGINE's EML macro-programming languages are examples. Since most spatial analysis software should also be able to read and write such files this provides a convenient and simple interface mechanism. While simple, this approach requires data buffering and, therefore, it is not normally used where dynamic linking is required. This approach is also only satisfactory for transferring relatively small quantities of simply structured data.

Database read/write

Greater flexibility and data management can be obtained by reading and writing data from and to a database table. All that is required is that both the GIS and the spatial analysis software can read/write data from/to a common database table. This approach is only really applicable to GIS software based on the hybrid or integrated design models. In the case of ARC/INFO, relational DBMSs (RDBMSs) supported via its Database Integrator link (Oracle, Ingres, Sybase, Informix, DB2, etc.) can be used. The strength of this approach is the flexibility it offers. Additional benefits include the fact that standard interfaces are often available between, say, ARC/INFO and Oracle, and a statistical analysis system and Oracle. The main disadvantages are that there is a processing overhead of storing and accessing data in an RDBMS, and that there is likely to be additional cost in purchasing the software for what amounts to a relatively simple application. This approach is also more time-consuming to set up than those discussed above.

Unix pipes and RPCs

A lower-level version of the programming language ASCII file read/write, is the transfer of data between GIS and spatial analysis software using a named Unix pipe or a Remote Procedure Call (RPC). A Unix pipe is a direct data-transfer mechanism between two applications, allowing rapid data transfer. Pipes can only operate on a single processor and can have limitations where the applications cannot be synchronized. Nevertheless, they are an effective method of transferring data at a slightly lower level than a programming language ASCII read/write.

RPCs are generic UNIX mechanisms for transferring data between processes running on one or more processors. The equivalent mechanism under Microsoft Windows is Dynamic Data Exchange (DDE). These are now standard, well-documented methods of transferring data dynamically between processes (e.g. a GIS process and a spatial-analysis-system process). DDE, in particular, is now widely supported by many application developers.

System programming

The final method to be discussed here involves the use of GIS software development library or object code routines. Most GIS software systems are written in the FORTRAN or C general-purpose 3GL languages. As such, they exist in the form of source, object and executable code. It is the executable code which is delivered to users. The object code can also be made available for users who wish to undertake specialist application development. Programmers can use the object code to create completely new applications incorporating some or all of the existing code. For example, a programmer could embed part of ARC/INFO's GIS functionality within a spatial analysis system, a new program could be written which embeds part of IMAGINE and part of a spatial analysis system, or part of a spatial analysis system could be embedded in ARC/INFO.

8.6.2 Classification and examples of combined 'Geographical Information System – Spatial Analysis Software' (GISSAS)

Over the past few years there have been many attempts to link GIS and spatial analysis software. Because most of those documented involve ARC/INFO in some form or another the discussion here is restricted to this GIS. This is not to say that others have not and could not be used. Rowlingson, *et al.* (1991) provide a useful review of early work. Goodchild *et al.* (1992) offer a classification of general GISSAS. It is the author's opinion, however, that this has some limitations in the context of the earlier discussion, and so an extended version of their scheme is proposed here. In this new scheme (Figure 8.4), GISSAS are classified at the highest level into Integrated and Hybrid systems. Integrated systems are those in which, as far as the user is concerned, the GISSAS is a seamless system. Integrated systems are subdivided on the basis of whether the GIS or the spatial analysis software is dominant. Hybrid systems are those which link GIS and spatial analysis software in a non-seamless fashion. These are sub-divided into close-coupled and loose-coupled systems.

Much of this work has been undertaken by researchers in universities and other government-funded research establishments and is characteristically of a prototype nature. Increasingly, however, GIS vendors and professional organizations are developing full, functional, industrial strength links (see Clarke *et al.*, Chapter 12, this volume).

In an integrated GISSAS type of system, spatial analysis software is dominant, and GIS object code is typically linked into spatial analysis software in the same way that ERDAS Inc. has embedded the ARC/INFO vector data model and handling routines in their IMAGINE software. As far as this author

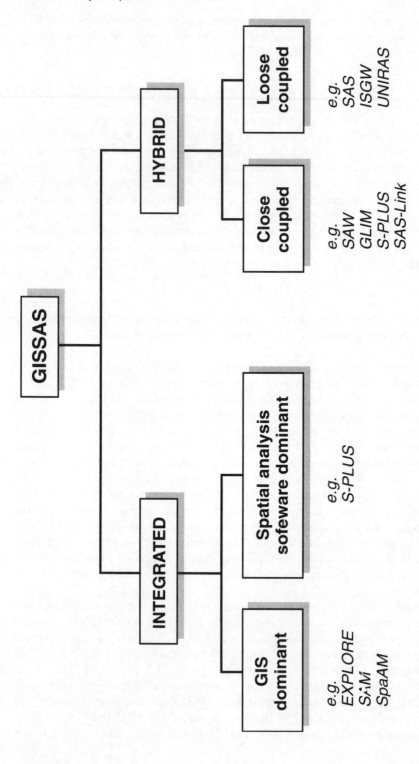

Fig. 8.4 A classification of Geographical Information System (GIS) – Spatial Analysis Software systems

is aware, there are no complete projects of this type, but some prototypes have been developed. Rowlingson and Diggle (1991), for example, sought to add spatial analytical functionality to S-PLUS (StatSci Corporation).

More usually, proprietary GIS have constituted the dominant components of GISSAS, with spatial analysis functionality then 'bolted on'. Majure and Cressie (1993), for example, have created EXPLORE, a prototype system for incorporating exploratory data analysis within ARC/INFO. This system is written in AML (ARC Macro Language) and can be used to identify anomalies in spatial and non-spatial data collected at fixed points over a period of time. Ding and Fotheringham (1992) have developed a statistical analysis module (SAM) that runs within ARC/INFO. This software is written in C and is called from AML. SAM can be used to estimate spatial autocorrelation statistics among other things. Carver (1991) has undertaken similar work and has created AMLs to add error-handling capabilities to ARC/INFO. Massie (1993) describes a solid-waste flow model developed using AML for Metro, Portland, Oregon. His system was coded in AML and SAS (SAS Corporation's Statistical Analysis System). A significant prototype GISSAS has also been developed by Jefferis (1993) called SpaAM. The system is designed to provide easier access to base ARC/INFO functionality, as well as custom procedures developed using AML. SpaAM contains functions for univariate statistics, exploratory data analysis, graphics and plotting, surface modelling, various utilities, visualization and display.

Hybrid GISSAS may comprise GIS and spatial analysis components which are either closely or loosely coupled. One of the first attempts to create a closely coupled GISSAS is described by Kehris (1990a; 1990b). He attempted to link ARC/INFO and GLIM. More recently, StatSci, the developers of S-PLUS, and ESRI have collaborated to couple closely ARC/INFO and S-PLUS. 'S-PLUS for ARC/INFO' provides interactive commands for transferring point, arc and polygon data sets, INFO data files, and GRIDs between ARC/INFO and S-PLUS. There is also a facility to issue any S-PLUS command from within ARC/INFO. A similar coupling exists between ARC/INFO and SAS in the form of SASLink, developed by American Management Systems Inc. SASLink provides ARC/INFO users with an interactive interface for automatically retrieving and creating SAS data files. SASLink adds six new commands to ARC/INFO. Nielsen *et al.* (1993) describe a numerical modelling package for simulation of oil and other chemical spills at sea. SAW is executed from the ARC/INFO interface, using information provided by the ARC/INFO environment (e.g. area of interest, specific location of the spills, etc.). SAW then takes over the display window and draws graphical simulations of spills. At the end, all output files containing oil-spill tracking data are converted into ARC/INFO coverages and can then be further processed or displayed using ARC/INFO.

Jackson's (undated) description of a loosely coupled link between SAS and ARC/INFO is an example of a loosely coupled GISSAS. This link comprises a collection of AML and spatial analysis software macros which convert data between the two systems via the ARC/INFO Export format. Davis and

Schwartz (1993) describe how to link ArcCAD to an Integrated Surface and Ground Water Model (ISGW). They output data from ArcCAD in ASCII format, post-process it using a FORTRAN program and then input it into their modelling software. Rowlingson *et al.* (1991) developed a link between ARC/INFO and UNIRAS which supports some exploratory data analysis. They developed a suite of FORTRAN programs allowing ARC/INFO and UNIRAS to share INFO files.

8.7 Discussion and conclusions

Interest in spatial analysis as a research topic has become enormous in recent years, with it featuring high on the research agenda of all major national GIS initiatives. Academics and researchers have been quick to extol the virtues of spatial analysis and deride the use of GIS for inventory applications. Spatial analysis is important for the same reasons that GIS is important: because many business and service planning problems require large quantities of spatial data for their solution. Many of these data have to be manipulated to make them compatible and then analysed to look for patterns and their underlying processes. Typical generic operations include: classification, interpolation, interaction, location–allocation and routing (travelling salesman). An important part of this analysis is the determination of the errors associated with any prediction. All of this activity must, however, be considered in the context of the fact that the vast majority of today's GIS applications basically involve simple mapping and query operations. As such projects mature and reach a higher level of sophistication, it is to be expected that more of them will become spatial analysis-orientated.

The brief review presented earlier, of the spatial analysis capabilities of two of the leading commercial GIS software systems, shows that many basic and some advanced spatial analysis capabilities have been added to commercial GIS software systems in recent years. Examination of the literature also highlights the fact that there have been a significant number of attempts to link GIS and spatial analysis software. Most of these have been research orientated prototypes, but there are examples of commercially orientated, industrial-strength systems (e.g. S-PLUS for ARC/INFO and SASLink).

Looking towards spatial analysis and GIS from a GIS vendor's perspective, it appears that there is a certain amount of academic elitism associated with the subject. Users and vendors are often led to believe that 'unless it's difficult it is not useful' and that 'data manipulation, description and display is not real spatial analysis'. In truth, data manipulation, description and visualization are probably the most widely used spatial analysis tools and the most revealing about data. Business and service planning users really want simple-to-use tools now, rather than an 'all singing all dancing' integrated spatial analysis information system in 10 years. The challenge for academic spatial analysts is to identify the most robust and reliable methods, to educate users about the benefits and pitfalls of using different methods, and to determine how to use spatial analysis simply (but with appropriate warnings about the degree of confidence that should be associated with results). A useful research project might determine the 20 per cent of the spatial analysis methods which will satisfy 80 per cent of user needs. Once the algorithms and applications have been determined, if there is user demand, it is relatively straightforward to add spatial analysis to commercial GIS software or to link commercial GIS and spatial analysis software.

9

Customized and proprietary GIS: past, present and future

Paul Cresswell

9.1 Almost everybody's doing it

Today everybody's doing it – well almost! Although an exaggeration, geodemographics and geographical analysis of markets is no longer the leap in the dark that it was in the early 1980s. Many business organizations are now employing the concepts of geodemographics and using GIS in many varying applications – from new site location analysis to merchandising and direct mail. With the increased range of applications to which geodemographics can lend itself, the market has grown considerably with new GIS software and databases to accompany these developments. This chapter will consider how proprietary GIS have developed and what their attributes are, and suggest some of the directions in which systems and information will develop in the future. On the basis of business experience, we will consider how proprietary systems have developed as the nature of the problems that businesses and service providers are faced with have changed.

Section 9.2 explores how proprietary GIS have evolved as the issues that businesses and services face have changed. An examination of the range of existing products is then undertaken in Section 9.3 and this is linked to some examples of GIS in action. Tailoring and customization of GIS is discussed in Section 9.4, with examples used to show how the complex issues that marketeers address can often be tackled within existing software and database frameworks. The role of GIS and market-analysis consultancy is an important area of involvement for the commercial arena and applications of this are highlighted in Section 9.5. Finally, in Section 9.6 a few thoughts are cast about how GIS might evolve and develop in the future and the impact that worldwide data exchange and software development platforms may have on steering the course for GIS.

The examples will largely be drawn from commercial applications. However, these examples are used first and foremost to demonstrate the capabilities of proprietary systems, rather than to argue from a not altogether independent position which is the best or worst. Many of the modular GIS available today have essentially the same functionality. It is in their design and ease of use that they differ most. The client examples used will be drawn from SPA's own client base. SPA's specialist areas are largely within the retailing arena, particularly in industry sectors such as public-house, restaurant and off-licence (liquor) retailing, and clothing and footwear.

The development of SPA Marketing Systems, a market-analysis consultancy based in Leamington Spa, UK, mirrors many of the trends in the market and as such serves as a useful illustration of how proprietary and customized GIS have been developed. The progression towards a strong consultancy-based organization has reflected the need for more focused GIS analysis as the issues that retailers and manufacturers face have changed since the 1980s.

9.1.1 From bureau to PC

SPA's development has given the company experience of both proprietary and customized geographical information systems. SPA was formed in 1984 by those responsible for setting up and running the retail planning service of an international market-analysis consultancy from its Leamington Spa branch office. But it had become increasingly frustrated in being able to offer clients only an expensive, inflexible, mainframe-based bureau service for analysing geographical information.

The opportunity was seen to offer a real alternative; a complete range of geographical analysis services on a client's own desktop personal computer (PC). These opportunities can be linked to the rapid changes in technology, as computer hardware costs plummeted and PC-based solutions became a very real proposition (Rhind 1991; Openshaw, Chapter 7, this volume; Maguire, Chapter 8, this volume). SPA was the first to put UK census data, lifestyle classifications and serious market-analysis software on desktop PCs. This was the beginning for us of the development of a whole suite of proprietary systems for analysing geographic information and applying these concepts effectively in business. When the bureau was king this was a revolution. Now it is a way of life.

9.1.2 Friendly software

SPA saw that the key to successful proprietary system development in these early days was user-friendly PC software for each of the key marketing tasks – area analysis, mapping, customer profiling, market modelling, leaflet distribution, branch location and segmentation. We all know how important geographical information is for making effective business decisions in today's competitive environment. However, geographic databases or locational information are of little use if the key patterns cannot be analysed effectively and efficiently.

9.1.3 Integration

Things have moved on dramatically since those early days. With more data available both from clients' own sources and from external data suppliers there is a far greater need for systems which are able to integrate diverse data-sets (for example, see Martin and Longley, Chapter 2, this volume). GIS cannot be exclusively preoccupied with the geographical dimension at the expense of other aspects of data structure. SPA have seen this trend emerge in the retail environment and SPA's company development is very much leading the way towards integrated analysis and reporting systems for retailers – systems which understand segmentation and turnover prediction and can link this with EPOS (electronic point-of-sale) or other branch-level data (see Birkin, Chapter 6, this volume, for a discussion of the linkage of lifestyle information).

9.1.4 Consultancy

In 1989 another company was added to the SPA group under the name of SPA Management Consultants. The consultancy was set up to provide analysis, advice and specialist software systems for a range of large retail clients. Since 1990, SPA Marketing Systems have positioned the business towards consultancy applications of GIS and away from GIS software development *per se*. This reflects the view that changes in the organization of the industry as a whole require applications to become more solution-driven rather than technology-led. In this respect the emphasis of our approach has been somewhat different from that described by Clarke and Clarke (Chapter 9, this volume), although many of the objectives remain essentially the same. SPA's main interests now focus on retail consultancy, modelling, geodemographics and market analysis. SPA Marketing Systems is a full agent of CCN Systems and can draw on MOSAIC and other CCN databases (see Birkin, Chapter 6, this volume) as well as on the software originally developed by SPA. We consider it important to offer not only state-of-the-art market-analysis systems, but the consultancy skills necessary to harness these to clients' specific needs – a vital partnership.

SPA's company development track highlights its move from largely being vendors of proprietary GIS towards a much heavier emphasis upon consultancy. Many retailers now have the tools to deal with their geographical analysis requirements, but need more help in applying them to wide-ranging commercial situations. At SPA we have therefore gained experience of both proprietary and customized solutions to geographically-related problems. Given the issues which retailers now face, the key is being able to respond quickly, open-mindedly and above all flexibly (Beaumont 1991; see also Clarke and Clarke's discussion of changes in the retail environment, Chapter 10, this volume).

9.2 Proprietary system development

9.2.1 Early development

Many tend to see GIS as self-contained environments which allow users to whizz between data and maps with a mouse and to derive answers to particular problems instantly – assuming they have asked the right questions! In fact some of the most widely used geographical analysis systems are modular. Early commercial systems to handle geographic data were somewhat rigid and structured in comparison.

In the mid-1980s few clients had vast reserves of their own data in a usable format. The early development of GIS was largely driven by the need to utilize and analyse available standard databases more effectively. In the UK these included the 1981 census and the associated classification systems such as early versions of CCN's MOSAIC and CACI's ACORN (Batey and Brown, Chapter 5, this volume). This was allied to marketing managers seeking to feel more in control of these data, and to interrogate such data quickly and easily in the privacy of their own offices.

Early software modules were designed to perform very specific jobs, such as to produce a census profile of any given area defined in terms of crude postal geography (pre-dating the postcode–enumeration district matches possible using 1991 UK Census data). SPA's early development of the ENVIRON area-analysis system was for this purpose.

Interest in such systems, certainly in SPA's early experience, came largely from the property and estate departments of large retailers, who used them almost exclusively to assess possible sites for new outlets using census data and proprietary area classifications such as MOSAIC and ACORN. In the early 1980s, use of the client's own customer information was fairly limited and the adoption of geodemographic and GIS techniques within marketing departments was in its infancy. The uses made of early systems were also a function of the economic situation. In the early to mid-1980s, 'economic boom' was rife. The requirement of early proprietary systems was very much geared towards the search for new locations for retail outlets rather than concentrating on the more marketing-led issues and problems facing retailers today.

9.2.2 Changing questions

During the 1980s the kinds of issues that organizations were having to deal with helped push the development of GIS forward. As more marketing applications opened up to retailers and competition for available sites became stronger, users of GIS were becoming more demanding in their requirements. The functionality and flexibility of such systems has improved as a consequence. The capabilities of systems in the early years were very much determined by the range of issues that retailers were addressing. At that time geodemographics, and geographic analysis more generally, was a very new concept. Retailers were feeling their way and dipping their toes in to test the water. An over-abundance of features and functions in early GIS may well have served to discourage their wider adoption.

However, as the premium new site locations were quickly devoured and the good times of the 1980s receded, the 'half decent' sites that were succeeding in the boom years were becoming more marginal, particularly as other retailers were getting in on the act. As other retailers were in the same position the market for selling retail space was limited: selling units was not always the answer. Such problems raised the prominence of a whole set of other related issues. This is not to say that these issues were not always around – just that they took on a greater significance in the late 1980s and early 1990s. The response has been to switch much more to the marketing uses of GIS and their wider dissemination into other departments within company organizations. With the changing nature of the problems that retailers were facing, marketing support and micro-marketing had now become a key objective and proprietary systems were developed to respond to these requirements. The range of issues that organizations now address using GIS are very wide-ranging. Rather than being used as a tactical tool as in many of the early applications, GIS is now beginning to be used as a strategic resource which can have an impact throughout the whole organization. Some of the typical issues that our own clients are faced with are:

1. Site location. Although GIS are still used for examining new site locations, they are increasingly being used to rationalize existing chains of outlets. Questions such as 'How do the catchments of our best performing branches differ from our worst?' and 'Do certain branches have overlapping catchment areas?' are now key areas of interest (see also Clarke and Clarke's experience of the television rental market, Chapter 10, this volume).
2. Acquisitions and mergers. A very important dimension in today's retailing environment is the opportunity to develop business by merger or acquisition. Assessing whether an opportunity could be successful or not often requires geographically-based techniques to examine such issues as catchment overlap and regional concentration (see also Clarke and Clarke, Chapter 10, this volume).
3. Merchandising. The application of GIS techniques to address retail merchandising issues is an interesting new development. For too long, many major retailers have based merchandising on past performance rather than on market potential. For example, a clothing shop that performed poorly in lingerie last year will have been given a more limited range and is therefore unlikely to do well this year either. This is the classic self-fulfilling prophecy. Merchandising based on shop performance is intended to reinforce success. When the Newtown branch does especially well in designer casuals it makes sense to push more stock its way in the following season. But, in the case of the lingerie store, such a strategy can perpetuate past mistakes. If sales are bad, surely the obvious course of action is to seek to find out 'why?'.

Retailers are now increasingly gearing what they stock in outlets to the local market. This is not only true of 'traditional' retailing. Industry sectors such as pubs and restaurants are now looking much more closely at what they offer to the local market, in an effort to increase turnover and get the best out of the sites that they already have in place. There is a whole new role for appropriate analytical techniques for segmenting and classifying branches or outlets into types in order to help merchandising departments make the most effective use of space.

4. Performance analysis. Given the slump in the property market, many retailers are trapped with a network of outlets that is difficult to rationalize. The development of GIS analytical functions that can help improve the productivity of existing outlets is a very important growth area in the consultancy industry. As such, there is an opportunity in the area of performance analysis for relevant research and the implementation of usable systems. As more customer data become available, many retailers are acquiring rich data sources feeding in from on-line computers. This breathes new life into statistical techniques of the early 1980s, such as regression analysis, that were theoretically sound but were of limited use because of the lack of appropriate data (but see Openshaw, Chapter 7, this volume). Appropriate techniques have always been available: it is the range of issues that retailers have to deal with in today's competitive environment that makes them more relevant than ever.

9.3 The range of existing GIS products

SPA's experience of GIS comes largely from the retailing sector. Most proprietary GIS to which retailers have access tend to be modular systems, purchased to do specific jobs within an easy-to-use PC-based software framework. Other GIS are very customized to specific applications, in terms of both the technical software used and the databases analysed (see Clarke and Clarke, Chapter 10, this volume). Customization of GIS can be evident at several levels. SPA particularly has an interest in customizing modular GIS in terms of the databases that they can interrogate and the applications for which they can be used, rather than on the technical issues of software development. We are also increasingly involved in data customization, putting clients' own data into GIS, for example, so that more general census data can be displayed alongside competitor locations and customer potentials. SPA's objectives are therefore very similar to those of Clarke and Clarke (see Chapter 10, this volume) in that the client's needs are the focus, so that analysis is directed towards strategic business needs, rather than towards technical niceties. Examples of this kind of customization are given later in the chapter.

For the more general marketing applications that retailers have, many prefer to invest in proprietary systems in the first instance. Several factors may contribute to this decision:

1. Cost. Because proprietary systems have often had a long development track and have been purchased by a number of clients, the cost of such systems can often make them more attractive than customized systems, which can involve heavy development time for one unique application. The typical cost of a proprietary PC-based GIS with standard databases such as the 1991 census would range from (UK) £20 000 to £25 000. Customized systems developed uniquely for a single client may run into hundreds of thousands of pounds.

2. Functionality. The best systems have an established client base and hence a large number of users. Shared experience and ideas from all business areas can therefore be fed into the development of such systems so that functionality is not limited to a rigid set of themes for one particular industry sector.

3. Ease of use. Large numbers of users dictate that the system should be easy to use and easy to support. An increasing number of proprietary systems today are very user-friendly to non-specialist users, as there is a move away from the need to understand programming languages and complicated operating systems. Much of the modular GIS software is menu-driven (such as CCN's ENVIRON geographic analysis system) and increasingly such software is being transferred into the Windows operating environment to link easily with other spreadsheet packages that many operators are now using to manipulate data. Ease of use is a very important consideration given the time constraints under which many clients operate.

4. Flexibility. Proprietary systems have to be flexible. They have to respond to new problems and to be able to exchange data with other systems in many file formats. As the nature of the problems that organizations face changes, so clients often wish to add in new modules or databases to perform specific tasks. A wide range of geographical databases and analytical functions that are now available can be 'plugged' into many proprietary software systems in order to open up a whole new range of applications. Again, development in computing environments such as Windows is relevant here as data exchange becomes more common.

However, the sophisticated proprietary systems that are now available can be of limited usefulness if they are not used properly or they do not address the clients' problems in a fast and efficient manner. In the market-analysis industry the systems that are available have to be seen as tools for reaching a desired solution for the client and not as an end in themselves. Naturally, when involved in site location and market planning the client needs to solve many problems and to make decisions quickly and efficiently. A successful proprietary GIS has to be able to respond to these needs.

Some of the most widely used systems are modular and perform a range of functions. The aim of this section is merely to give a flavour of the kinds of application and functionality that current proprietary systems can offer, picking several key GIS applications and highlighting analysis issues that are commonly addressed by such systems.

9.3.1 Postal geography and proprietary systems

Before discussing some of the features of proprietary GIS, it is worth highlighting that the development of systems has also gone hand in hand with the establishment of postal geography as a standard set of building blocks for the marketeer. Without a unit of geography which was familiar and linked to users' real-life locations, it is doubtful whether the widespread usage of proprietary GIS would ever have grown in the way it did. This is demonstrated by the unwieldy nature of early UK GIS, which were based on the census enumeration district as the unit-level building block (see Raper *et al.* 1992).

For GIS, postcodes/zipcodes and the whole postal system is nothing more than a means to an end. The key feature of the postcode is that it can be used as a label for an area, and only if this area can be 'typed' or categorized in marketing terms is this of real value. Unit postcodes have obvious advantages for marketeers because they typically comprise addresses and can be aggregated to create almost any larger area. In addition they are familiar to the general public and much marketing information from surveys and questionnaires is collected with a postcode to categorize the respondent.

Postcodes are therefore fundamental geographical building blocks. For market analysis, they can describe and locate the target market of both existing and potential customers. Using the postcode, proprietary GIS are able to incorporate within their structure a range of standard geographies, such as postal areas, local authorities, economic regions, media geographies and retail catchment areas. Most proprietary systems use postal geography as a way of building catchments, with the postal sector being the most common building block in the majority of retail applications. The systems described below all use postal geography as the key representation of location.

9.3.2 Area analysis

Systems such as CCN's ENVIRON and CACI's InSite geographic database managers are effectively able to let the user analyse geographic databases, built out of various hierarchies of postal geography. Databases such as the census and area classifications such as MOSAIC can be analysed for any geographical area that the user chooses. Modelled potential databases are also available, such as SPA's TASTES database for the drinks and leisure industry. A list of where these databases may be obtained is included in Appendix C. Databases of where key competitors are located are also becoming very important. Common features of such area analysis systems are:

1. Area profiling. Users can define catchment areas around particular locations in a variety of ways.

- Lists of eligible postcode areas, districts or sectors: see Raper *et al.* (1992) for an authoritative discussion of postcode geography.

- Circular areas (e.g. within a five-mile radius of a planned shopping centre).
- Polygons (the area within a boundary drawn on a map).
- Drive time areas (e.g. 30 minutes' drive from a regional shopping centre).

For example, a zone can be defined based on the drive time from a point, or a circular radius. The operator can even import grid references from a shape drawn on a detailed map. Any catchment can be compared with any other catchment, and the operator can, for example, look at the areas of catchment overlap or examine the demographic profile of areas which are not served by an existing outlet network. Typical questions which area-analysis systems are able to address are:

- What is the population profile within 20 minutes' drive time of Birmingham and does it have concentrations of my kinds of customers?
- How does my York branch catchment compare with my Birmingham branch?
- Are there any geographical patterns which might help to explain the difference in performance between my different outlets?
- What are the implications for merchandising? For example, is Perth better for stocking more jeans than suits?

To illustrate how area profiling works, Table 9.1 shows information from the 1991 UK Census for a 15-minute drive time around Kingston-upon-Hull. The totals for Hull are compared with the figures for Great Britain (GB, i.e. the UK minus Northern Ireland) as a whole and the bar chart on the far side of the report shows whether the area is above or below the national average for each variable (100 = GB average). Using proprietary systems these reports can be generated and printed in a matter of minutes, in order that clients may 'wander' around their branch network without ever leaving their desks! However, there are problems with a naive approach to drive time analysis: see Birkin, Chapter 6, this volume.

Area analysis systems are operational in many countries. CCN have developed applications of ENVIRON for many European countries and the American market has long been a centre for such analysis, with use of census data being in the public domain and therefore free. Some of the databases that are available are described later in this chapter and contact addresses of some relevant vendors appear in Appendix C.

2. Catchment rankings. Clients often use area-analysis systems to produce a ranking of a set of catchment areas on a particular variable. This could be a list of their own outlets or a list of potential new sites. Such techniques are used increasingly for analysing competitor sites or preparing the ground for an acquisition or merger. For example, an application might consider which existing pub (bar) sites in the client's portfolio have the greatest catchment area potential for offering quality food, using such analyses.

3. Local marketing. The concept of marketing within a particular store's catchment area is widely used by a variety of retailers. Ranking of postal sectors on a relevant indicator of patronage within a catchment is often used as a basis for a leaflet drop or for other promotional activity. Given increasingly flexible printing technologies, the contents of the leaflet are often geared towards the characteristics of the local catchment area. Important also is the use of GIS for local media advertising and joint promotions with other retailers. For example, CCN's MOSAIC classification is widely used by the major newspaper groups in conjunction with the ENVIRON geographic analysis software in every element of newspaper marketing from reader profiling and sales-performance analysis to advertising-sales analysis. It is now even possible to link area analysis to direct-marketing databases. For example, a new cinema opening in a town could define its likely catchment area, and then by linking the catchment definition into a source such as the UK Post Office's Postal Address File (PAF) a list of address ranges for direct mail promotion can be derived.

9.3.3 Mapping systems

One of the most noticeable changes of emphasis from early proprietary systems is the increased prominence of graphical forms of displaying results compared with the more traditional tabular forms of report. Many area-analysis systems now routinely link into mapping and graphical systems which allow display of the salient geographical information. This routine use of graphical display will be very important in future GIS development as analysis becomes driven by maps and graphs.

Several general mapping systems are available. A distinction is often made between two types of system:

1. Area mapping products. Products such as CCN's MACROMAP produce colour-shaded maps from data which are often held at postal-sector level. MACROMAP links directly with the ENVIRON area-analysis system in order that clients can map catchment areas and portray any of the data that are available in their database. Such maps are often used by users of the system when presenting results to senior management, who now regard map display and production as an integral part of any project that uses geographical data. An example of such a postal-sector map is shown in Plate 9.1, depicting the level of car ownership around the Bristol area.

2. Point mapping products. It is increasingly the case that choropleth (shaded-area) maps are being superseded by map point data from grid references. As more customer data become available these systems are very useful for mapping customer distribution around a store, depicting the make up of an area in terms of a geodemographic classification, or mapping the location of the analyst's stores or those of their competitors. With intelligent use of colour

Table 9.1 Area profiling of 15-minute drive time catchment of Kingston upon Hull

---

1991 CENSUS REPORT

Title: STATUS\AGE\BIRTHPLACE\CLASS   Client: Example
Target Zone: 15 MINS HULL   Base Zone: GB (GB)

	- - Target Zone - -		- - Base Zone - -	
	COUNT	RATIO	COUNT	RATIO
RESIDENT POPULATION				
All Residents	282 471	100.0%	54 632 124	100.0%
Males	136 706	48.4%	26 431 834	48.4%
Females	145 765	51.6%	28 200 440	51.6%
MARITAL STATUS				
All Residents 16+	221 105	100.0%	43 624 808	100.0%
Married	123 009	55.6%	25 547 942	58.6%
Single	98 096	44.4%	18 075 882	41.4%
Single parents	5 887	2.7%	823 658	1.9%
BIRTHPLACE				
All Residents	282 471	100.0%	54 632 124	100.0%
United Kingdom & Eire	277 352	98.2%	51 519 480	94.3%
Old Commonwealth	372	0.1%	174 952	0.3%
New Commonwealth	1 901	0.7%	1 673 434	3.1%
Rest of the World	2 846	1.0%	1 264 215	2.3%
AGE STRUCTURE				
All Residents	282 471	100.0%	54 632 124	100.0%
0 to 4	21 526	7.6%	3 625 351	6.6%
5 to 14	36 297	12.8%	6 728 269	12.3%
15 to 24	42 327	15.0%	7 563 979	13.8%
25 to 44	81 437	28.8%	15 943 323	29.2%
45 to 64	57 554	20.4%	12 004 244	22.0%
65 to 74	24 965	8.8%	4 930 035	9.0%
75 and over	18 365	6.5%	3 833 841	7.0%
All infants and children	61 451	21.8%	11 003 031	20.1%
Adults of working age	170 240	60.3%	33 402 912	61.1%
Adults of retirement age	50 780	18.0%	10 225 083	18.7%
SOCIAL CLASS				
Private households	114 190	100.0%	21 874 098	100.0%
Total household heads	66 047	57.8%	13 493 754	61.7%
Class 1 Professional etc	2 627	4.0%	901 881	6.7%
Class 2 Managerial & tec.	13 738	20.8%	4 123 373	30.6%
Class 3N Skilled non-man.	8 248	12.5%	1 828 742	13.6%
Class 3M Skilled manual	20 944	31.7%	3 569 618	26.5%
Class 4 Semi skilled	11 461	17.4%	1 839 691	13.6%
Class 5 Unskilled	5 082	7.7%	632 861	4.7%

Source: Office of Population Censuses and Surveys 1991.

- - - - - Target/Base - - - -	-0	50	100	150	200	250	300
PENETRATION INDEX	|---|---|---|---|---|---|---|---|---|---|---|						

0.005 17	100	
0.005 17	100	
0.005 17	100	
0.005 07	100	
0.004 81	95	■
0.005 43	107	■
0.007 15	141	■■■■■
0.005 17	100	
0.005 38	104	■
0.002 13	41	■■■■■■■
0.001 14	22	■■■■■■■■■■
0.002 25	44	■■■■■■
0.005 17	100	
0.005 94	115	■■■
0.005 39	104	■
0.005 60	108	■■
0.005 11	99	
0.004 79	93	■
0.005 06	98	
0.004 79	93	■
0.005 58	108	■■
0.005 10	99	
0.004 97	96	■
0.005 22	100	
0.004 89	94	■
0.002 91	60	■■■■■
0.003 33	68	■■■
0.004 51	92	■
0.005 87	120	■■■
0.006 23	127	■■■■
0.008 03	164	■■■■■■■■

coding, point maps of customer locations relative to the location of the outlets at which they shop reveals much about the catchment area of individual stores. They also neatly circumvent the modifiable areal unit problem (MAUP) discussed by Openshaw (1984b; see also Martin and Longley, Chapter 2, this volume).

One of the important current issues for retailers is linked to the threat of competition not only from key competitors, but from cannibalization of trading areas by outlets from within the same branch network. Proprietary mapping systems allow clients to investigate these kinds of problems using desktop PCs. The example map in Plate 9.2 shows how point mapping can be used effectively to identify areas of overlap between the customers of several stores in the chain, and to pinpoint gaps in the market. Point maps relate to the location of the population much more realistically than area maps, and can be linked directly to road network databases.

9.3.4 Customer catchment analysis and profiling

As the benefits of collecting more customer information are gradually realized, the potential for systems that analyse customer data as well as external data sources has dramatically increased. Increased use of customer surveys, electronic point-of-sale (EPOS) data (see Guy 1994), and the collection of store-card and charge-card information, has led to marketing applications that rely on gaining more knowledge about general customer/client bases, identifying where big spenders/heavy users are concentrated, and assessing related lifestyle/patronage patterns (see Birkin, Chapter 6, this volume). To make use of these data, proprietary systems have developed modules that can shed much light on customer behaviour and discover where more potential can be tapped.

1. Catchment analysis. Customer postcodes are now routinely collected, and so clients can now use their own data directly in order to analyse the catchment areas of their stores. Several proprietary systems on the market allow the analyst to define the catchment area of stores using customer postcodes and the code of the branch they visit. Accurate catchment areas can be defined for each branch in terms of drive times. Areas of catchment overlap can be identified and the analyst can quantify the average drive times needed to shop around different types of outlet. There are limitations associated with such a simplistic analysis of catchment areas, such as the limited account taken of intervening opportunities such as competitor stores (see Clarke and Clarke, Chapter 10, this volume). However, many clients have neither the budgets nor the more extensive data sources that are necessary to develop more specialist solutions, and so the ability of proprietary GIS to at least offer ways of producing benchmark statistics is invaluable to business and service planning.

2. Customer profiling. Many organizations are still unable to describe who their customers are in any systematic way. However, a common use of proprietary GIS is now to enable customer profiling using some of the standard area classifications such as CCN's MOSAIC or CACI's ACORN. The analyst can match a list of postcodes against an area-profiling system such as MOSAIC or ACORN in order to derive a profile of different kinds of customers. The example in Figure 9.1 shows a customer profile based on MOSAIC groups. Three groups are prominent in this example profile: 'High income families', 'Suburban semis' and 'Country dwellers'.

Using this knowledge one can use area-analysis systems to find more of the same types of customers or to target local marketing to specific sub-groups of the catchment area which would be more responsive to a particular offering. Provided the sample size is large enough the analyst can also generate separate profiles for individual products which show the sub-groups that are most likely to buy the product or to spend the most.

The knowledge gained from customer-profiling analysis may also be used to build a geographical potential model. Such an approach to modelling is based upon modelling distributions based on areal data, rather than on the inferential models present in approaches such as those based upon spatial interaction. By matching a customer profile with the profile of an area on the ground, models

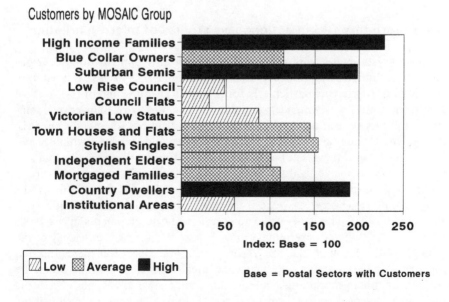

Fig. 9.1 Customer profiling using the MOSAIC geodemographic system

can be built at small area level to provide a bench-mark as to whether the area is good or bad for a particular product or retail chain. The same geo-demographic modelling principle can be applied to industry market-research surveys such as the Target Group Index, the TMS Clothing and Footwear Survey, Stats MR and PAS drink surveys, and CMT's National Shoppers Survey (see Appendix C for address details). Even if clients have no customer data of their own there is the capability for proprietary systems to contain market-relevant potential models.

9.3.5 Classification systems

As the range of issues that retailers are addressing has changed, so classifications of branches into types for specific marketing applications have also developed. Although such work is still largely consultancy-based, some clients do have proprietary systems which allow catchment area classifications to be carried out on their own PCs. Such systems examine the profile of each catchment area (perhaps the modelled propensities to purchase a range of key products and services) and classify the catchment according to the style of outlet that is most appropriate there. As merchandising and style of outlet become weapons which retailers can use to try to win back the customer and make their offering more attractive than that of their competitors, the future holds much potential for further development of these systems for wide dissemination.

9.4 Proprietary GIS in action: case studies of micro-marketing

Without crystal-ball-gazing for specific and perhaps unpredictable developments, one thing that we can nevertheless say about the 1990s is that it will be a decade of great change, particularly for retailers. The 'greying' of the Western European and North American populations is a widely-recognized phenomenon, and this general ageing will generate new demands for market analysis. As affluent societies get older, so there will be demands for higher quality and better service, and there will be a sharper marketing focus on small groups of individuals and the individual town, rather than analysis aimed at broad groups of consumers or large geographical regions. That is to say, this will be an extension to the previous trend from mass- to niche-marketing identified by Beaumont (1991). Thus one of the most significant developments in the 1990s will be the local dimension to marketing. Whilst retaining the efficiency of the multiple-outlet formula, there is likely to be more concession merchandising together with design tailored to the local context. This suggests a growth in the concept of 'micro-marketing', determining which sorts of people live in the local area, where they live, and what sorts of lifestyles and aspirations they have. There will still be very profitable niches to be exploited in the high street, but precise targeting of consumers is likely to hold the key to retail success in the late 1990s.

It is in this context that it is useful to look at selected case studies which illustrate how the use of GIS software and databases can help organizations address local marketing issues. The examples are typical of the kinds of issues for which marketeers are able to use proprietary GIS systems, without resorting to very expensive customized solutions. The targeting applications of market analysis and geodemographic databases are just part of the applications of geographical data, but are set to be increasingly important in the 1990s as retailers get closer to the local market.

Every year countless millions are lost by blanket mailing of leaflets to areas which are never very likely to respond to their message. The following case studies will show how market analysis can help to cut costs and increase efficiency. A further aspect of local marketing that will be discussed here is the maximizing of media potential and its role in micro-marketing in the 1990s.

9.4.1 Promoting the petroleum!

As a result of careful planning, the new petrol (gas) station sits by the roadside with its gleaming new pumps and forecourt. But whatever the potential that exists in the surrounding catchment area it will not be very successful as a business proposition if the right kinds of people know nothing about it.

This was the problem when a new petrol site was opened in Sale, Cheshire, UK. In order to advertise its facilities it was decided that there was a budget for about 50 000 leaflets to be dropped to households within the surrounding 15-minute drive time catchment area. However, 50 000 leaflets delivered to the wrong households would not only be very costly in terms of producing the leaflets but would also bring little extra revenue through the pumps. The objective was thus to ascertain the best postal sectors within the 15-minute drive time for door drop activity and to maximize the return from them.

The first step was to find out who the customers of the new petrol station were likely to be and then to target areas with high proportions of these key groups (in a similar fashion to the hypothetical pizza-delivery problem described by Martin and Longley, Chapter 2, this volume). However, as the garage was a new site, there was, as yet, no information available about the customers who would be likely to visit it. A MOSAIC profile was therefore constructed to reflect which of the MOSAIC types were most likely to have high propensities to visit the garage. High users of petrol would obviously be important. Using MOSAIC coded information from the Target Group Index Survey we could obtain a MOSAIC profile of likely high petrol users. The Target Group Index is a market research survey completed by over 25 000 respondents each year. The survey contains questions that ask specifically about petrol usage. By linking a respondent's postcode to a classification system such as MOSAIC, a profile can be constructed of the MOSAIC areal types where petrol consumption is highest.

SPA then produced an analysis which, with information from the MOSAIC profile, models a potential estimate for every postal sector within the 15-minute catchment definition and compares it with the number of households in each postal sector. This penetration rate is then simply a measure of the sales the client would be likely to achieve per household in each postal sector. If the clients are dropping a limited number of leaflets they obviously want to target those postal sectors where the adoption rate per 1000 leaflets is likely to be highest. In this example, 18 postal sectors out of a catchment area of 95 sectors were used by the leaflet distribution company to target their field force. By concentrating the campaign to those 50 000 households where the sales/household ratio is highest the client could get more cars to the pumps at a fraction of the cost of dropping leaflets to every household within the whole 15-minute catchment area. What this example demonstrates is that knowledge of the client's local market can allow the client to increase efficiency and achieve mailing or leaflet distribution which is both cheaper and more direct whilst still attracting people through the doors. (However, see Birkin, Chapter 6, this volume, for the caveats to using TGI and other survey information in this way.)

9.4.2 The manufacturer's response

Manufacturers are now embracing GIS techniques as they fight for every available metre of retail space. Micro-marketing is a concept which is equally applicable to manufacturers as well as retailers. The move away from the mega brand of the 1970s towards much more tightly defined products has important consequences for brand marketing in the 1990s. The 1980s can be seen as a period in which the balance between the manufacturer and the retailer was tipped increasingly towards the retailer. The 1990s may be a period in which manufacturers reassert some of their former strength, and geographic information will be a key tool in any such shift.

We all know that some brands sell better in some parts of a country than others. But, where are these areas and what sorts of people are buying which products? The message is that to influence their retailers, manufacturers must know their end-users. By using geodemographics, manufacturers too can have more accurate, more understandable, and more actionable customer profiles. They, too, can estimate sales potential with computer models, promote in the right areas and select the right media. Additionally they can build efficient direct mailing lists which are targeted to people who live in the right places. To illustrate the use of GIS techniques in the manufacturing sector, this section describes the case of a large manufacturer making household cleaning products, who was about to launch a new product.

The prime marketing objective for the manufacturer was to define and contact its customers by designing and implementing a questionnaire. The methodology began with a mail drop. There was a degree of rough targeting at this stage as the consultants did not want to redefine the response by only

Customized and proprietary GIS

sending the questionnaire to certain types of people. So, an area was chosen – the TVS television region in the South of England; the more deprived of the council housing areas were excluded and 300 000 questionnaires were dropped, again targeted by an analysis produced using proprietary software.

To determine what the base for comparison should be, a MOSAIC profile of the whole drop area was produced by the consultants. The 'base' category against which the targeted area was to be compared was in this case the television area rather than Great Britain as a whole. The questionnaire itself had a large number of questions about the lifestyles and buying habits of respondents. It was supplied with a 25p coupon to encourage trial of the product, with a promise of further savings if the questionnaire was completed and returned.

Response rates were very encouraging, with over 7 per cent of the 22 000 questionnaires returned. The responses were built into a database that SPA then processed. Computer software checked the postcodes supplied and then added a MOSAIC code onto each customer record. The client could then begin to produce profiles of regular and occasional users of the client's own brands and those of competitors, albeit being wary of introducing the possibility of ecological fallacies common when studying spatial patterns based on aggregates rather than on individuals (see Martin and Longley, Chapter 2, this volume). All of this information is now being used by the client's 'above the line agency' to help target their television commercials and press advertisements.

By drawing on the results of this exercise we were also able to target a door drop of 1.5 million leaflets and product samples to help launch a new product in the range. As it was a new product, no information existed about which MOSAIC types the customers would come from. This profile was built up by looking at similar products and seeing which factors were associated with their popularity. For example, 85 per cent of users of a similar product had children under the age of 4. Using this MOSAIC profile in the manner described above, SPA was then able to produce a list of postal sectors consisting of the 1.5 million households who would be most likely to buy the new product.

Proprietary GIS have also enabled manufacturers to reinstate ailing products using market-analysis tools and geodemographic software. Take the case of a fizzy drink, made by a major soft drinks manufacturer. For a number of years the particular product had been slipping in popularity. The manufacturer wanted to build awareness amongst consumers of the drink and realign the brand to be more appealing to today's family. They decided on the family as a focus after a piece of research showed clearly that there were two key purchasers of fizzy drinks – schoolchildren and mothers with young families. Using the variables that make up each MOSAIC type in conjunction with the research data, a MOSAIC profile of potential purchasers was constructed. This was then used to target a 4 million leaflet distribution. By using this method of market analysis the manufacturer targeted their leaflet distribution to key households and consumers more accurately than traditional techniques would have allowed.

9.4.3 Maximizing media potential

One of the problems that many marketing managers are keen to address at the local level is how to place their advertising so that it reaches the maximum number of customers. Which newspapers or magazines have the highest readership penetration of their sorts of buyer? How can best use be made of the advertising budget?

On a practical level, market analysis and geodemographics can now address such problems as drawing up a media schedule. For example, a Media Potential Analysis can produce a practical solution to these problems. This technique compares a MOSAIC profile of the client's target market (existing customers, existing branch catchment profile, or profiles of products from a survey such as Target Group Index, the TMS Clothing and Footwear Survey, or the Financial Research Services Survey) with the MOSAIC profiles of the readers of up to 250 newspaper and magazine titles from the National Readership Survey (NRS). Reports from this analysis rank groups of titles, giving for each one the estimated number of potential customers and the percentage that they represent of the title's total readership.

Consider a company selling skiing holidays. Where do we advertise holidays in the snow? In terms of readership penetration, the result is fairly obvious; the quality end of the market scores highest, with the UK *Daily Telegraph* and *Times* broadsheets ranking highly. But ranking the titles by numbers of potential skiers shows a very different picture. In fact, the message, if you are looking for sheer volume potential could be 'Head for the (tabloid) *Sun* if you want snow'!

Recent work at SPA has focused on using GIS techniques for the right media selection for a large up-market clothing company. As readers can be linked to area classifications, it is possible to produce areal models which can link an NRS profile of readers of a particular title to the areas in which they live. This sort of analysis can help with all sorts of media schedule decisions: where should we advertise our new range of in-car CD players – *Car*, *Fast Lane* or *What Car*? Are we missing an opportunity in *Vogue* or ignoring the potential of *Yachting World*? What regional newspapers are best for advertising the opening of my new toy shop? What are the best radio stations on which to announce the new range of clothing in my menswear stores? This brings GIS consultants into close contact with direct mail and advertising specialists, blending together a unique range of skills into an integrated solution for the client.

All of these questions are closely linked with the concept of geodemographics at the micro-level. It can now be demonstrated that a different media schedule would reach the same percentage of a target audience at sometimes up to half the original cost. There is nothing particularly magical about all this. What it does mean is that geodemographics now allow clients to reach the target market with much lower advertising spends than they would otherwise have been able to do without such techniques. The 1990s will undoubtedly see a still sharper focusing upon these issues.

9.5 Tailoring and customization of GIS

Any good proprietary GIS will have a range of different functions and capabilities. However, if these are not harnessed in the right way, the benefits of using such tools are greatly reduced. There is a very definite opportunity to offer consultancy to clients who are using proprietary systems in order that they get the best out of what is after all a very substantial purchase. This involves not so much educating the user in which keys to press, but suggesting new uses for existing tools and helping the client to formulate structured approaches to what may be very complex problems. This is one area of business that my own consultancy has developed, using existing proprietary system platforms but emphasizing the consultancy approach to using such software, and deploying our skills as analysts.

If an organization needs customized development to make GIS work in its environment, it should see the vendor and/or consultant as a development partner. For this to work, a high level of trust needs to be developed on both sides. Many organizations still see GIS as a tactical toolkit rather than a strategic resource that can extend its influence throughout many facets of an organization's structure. By keeping close to the end-user client, development ideas are directly influenced by the current issues facing the real world. There are several roles for the analysis industry that are particularly relevant for clients with proprietary systems.

9.5.1 New uses

When working closely together with a client, problems often arise which do not immediately seem to be addressed within current system structures. However, the answer often lies in imaginative ways of using proprietary systems. For example, an increasingly important issue for retailers who have high numbers of outlets is not only how closely competitors overlap with their catchment areas, but also how one outlet in the chain is cannibalized by other outlets within the same chain. These questions are particularly pertinent for issues of rationalization, merchandising or local store promotion. In a recent project we were able to handle information from a customer survey in a geographical way and to manipulate the data into a format which could be analysed through the standard ENVIRON area-analysis system. Rather than supplying a census profile on the screen, the client can now see, for any given catchment area, which of their other branches draws custom from that catchment, as well as whether and to what degree this is a problem to be accommodated in planned promotions. These kinds of database have also been used to analyse media catchments in order that retailers planning local media campaigns may determine which newspaper catchments cover any given catchment area and which cover the most potential customers.

9.5.2 Relevant databases

One of the most important roles of tailoring and customization in GIS is in building, modelling and operationalizing relevant geographic databases for specific clients. Rather than clients requiring 'blanket' census databases, as in the early days of proprietary systems, there is now a greater need for information from other sources to be built into geographic databases. Clients often wish to link census data with an area classification system, alongside actual and modelled customer data. Of increasing interest is the ability to add data on competitors into area analysis systems, using data from sources such as Retail Locations (see Appendix C). The impact of competition, within, say, a one-mile radius of a client's store can be examined. This adds another dimension to otherwise purely demographic analysis and brings into the mix some of the other factors that we all know affect store turnover.

At no stage when using proprietary systems must the client or the consultant lose sight of the quality of the data that are being used for analysis. A GIS can have as many 'knobs and whistles' as possible but if the data are poor the key *raison d'être* for investing in such a system is undermined.

9.5.3 Models on the catwalk: clothing and footwear expenditure

As an example of the kinds of geographical databases that can be built to address specific market sectors, this section looks at the development of a database for the clothing and footwear industry – a database which can be analysed using standard proprietary GIS.

In conjunction with the TMS Partnership (TMS), a leading specialist in the provision of high-quality information relevant to the clothing, footwear and accessories industry, SPA Marketing Systems have developed the UK 'Clothing and Footwear Databases', offering geographical information about a whole range of clothing, footwear and accessories. The information relates directly to the needs of market planners, manufacturers and retailers in the clothing and footwear industry. Such a database will allow clients to compare, for example, the catchment areas of Nottingham and Northampton, or Bristol and Brighton. This information can then be used to investigate prospective sites and target areas for expansion or rationalization with a certainty lacking in some of the more traditional market-analysis methods.

The Clothing and Footwear Databases are extremely comprehensive, offering detailed market size estimates for many products and covering all types of consumer market. The variables in the Clothing Databases are modelled potential-expenditure estimates in thousands of pounds, based on information collected in the annual Clothing and Footwear Survey carried out by Textile Market Studies (TMS). The TMS Clothing and Footwear Survey is a survey of about 80 000 consumers and is designed to measure the markets for all of the main items of outer-wear, underwear, footwear and accessories for adults and children. The TMS survey is postcoded and can therefore be linked to CCN's MOSAIC lifestyle classification. Each survey respondent can be allocated to a MOSAIC type. A MOSAIC profile can then be built up from the data in the survey. For example, a profile of the sorts of people who spend high amounts on footwear might be constructed.

Linkage of up-to-date market research information with the powerful MOSAIC lifestyle classification provides a basis for modelling the potential for a whole range of clothing and footwear products, based on the notion that people who live in similar areas tend to have similar lifestyles and buy broadly the same kinds of goods and services. By classifying areas into types, therefore, small area classifications identify major lifestyle categories, which are invaluable in assessing area potential. The MOSAIC profile shows which of the MOSAIC types contain the most expenditure, and which contain the least. So, if the clients know the make-up of an area's population by MOSAIC type, it is fairly easy to estimate the total expenditure that is likely to be generated by that area.

Resulting expenditures by postal sector are thus modelled potentials. It must be understood that they are not figures showing the actual level of spend on each product in an area, but an estimate of what the level might be expected to be given the sorts of people, in terms of MOSAIC lifestyle types, that live in those areas. For example, the model can suggest that the level of expenditure on men's footwear is high compared with the national average in Weston-super-Mare, even if there is no footwear shop in the town, the implication being that the potential exists for a shop to be located here. This makes the databases an important planning tool in deciding how much more potential for a particular product exists in an area and how best to allocate resources for achieving the client's desired targets.

There are obviously far more complex techniques available to analyse past performance and to predict future performances, such as gravity modelling, which could take into account the effect of competition (see Clarke and Clarke, Chapter 10, this volume). However, some of these methods need more data, which are often not available in the volume needed for more complex methods to be successful. In contrast, use of relatively simple area modelling techniques, particularly in conjunction with good classification systems, can identify key points which were previously hidden in a sea of data.

The Clothing and Footwear Databases can be accessed by the ENVIRON geographic database manager, allowing the client to interrogate the data and ask his or her own questions. Clothing and footwear profile reports of any area in the country, defined by postal sector geography, or even drive times, can easily be derived. The results can then be mapped using powerful mapping software.

Having defined a catchment area, the client may wish to explore its composition, perhaps with a view to some micro-marketing. Is the client advertising a new sportswear shop by distributing leaflets to households within a 15-minute drive time of the store, promoting the new service? What are the postal sectors with the highest potential expenditure on sportswear? Which sectors have the highest average spend on tracksuits, swim-wear or other sportswear? These concepts apply equally as well to clothing or footwear manufacturers and suppliers. Geographical databases can target mailings for the client's product in those areas, around the stores they supply, which represent the people who will be most likely to be interested in the product. If one is distributing leaflets, software can be used to target a distribution service at the sectors within the catchment area that will be most rewarding: if the target group is, say, expenditure on women's clothing, it makes sense and saves money to direct leaflets at those sectors with the highest expenditure in this target group.

Perhaps some of the information held at individual postal-sector level is too detailed for a first filtering of suitable locations from unsuitable ones. At the initial planning stage of any project clients often want only an overview or a feel for which location or towns might offer the right ingredients for a new store location, a franchise area, an improved branch product mix or, for a manufacturer, deciding who should supply a product to its end-users. Again market-analysis tools can help. Clients can choose their target group – this might be expenditure on women's underwear, men's accessories or children's jeans. They can then report on the towns that offer the highest potential for the particular product or service.

By asking the right kinds of question, a geographic database in action can be an extremely valuable planning tool, as this case study has shown. The value is increased further when one considers that such databases can be interrogated by marketing managers on a desktop PC, within a matter of minutes.

9.5.4 MOSAIC in education

As an example of new uses of GIS software and databases, consider the application of MOSAIC in education planning. UK Local Education Authorities (LEAs) are set to become key users of MOSAIC, the lifestyle classification developed for consumer marketing. The UK Education Reform Act (ERA) has moved many schools into a competitive market for pupils, where examination results are likely to be a key factor in parental choice. Yet exam performance is strongly affected by social background. So LEAs are required to produce formulae for the distribution of funds to schools according to social and educational need.

Free school meals have long been used as a measure of social disadvantage (since they are offered to the children of low-income parents); and schools with high proportions of pupils eligible for free meals are often given extra funding. Working with Loughborough University's (UK) Department of Education, SPA Marketing Systems have devised a better approach, producing a tailor-made MOSAIC analysis for Bradford LEA. Some 65 per cent of the variation in secondary schools' exam performances is explained by MOSAIC, compared with 40 per cent by free school meals. For the technique to work, all the system needs is the full postcode of each of the pupils of each school, together with data on exam or Standard Attainment Test (SAT) results. This type of analysis provides a more objective way for LEAs to begin to look at fund allocation. During the next decade we shall see such GIS techniques applied to other increasingly varied areas in service planning.

9.5.5 Data consultancy

There is a very important role for market analysis companies to play in highlighting differences between data-sets, invoking very rigid quality control standards and making clients aware of the strengths and weaknesses of the various data sources available. This is not only true for externally available data-sets but also for clients' own data. A vast amount of our time is spent checking the validity of data that clients may have collected from surveys and suchlike before deciding whether or not it is usable. As data-sets become more rich and varied this problem is unlikely to diminish. It becomes more important than ever to quality-check data collection, in order to ensure appropriate database design and to assess the appropriateness of each data source for the purpose for which it is to be used. The progression from data to information to decision-making and finally to success has to begin with good data.

9.5.6 UK census data, Euripides and specialization

The release of 1991 UK Census data at small area level to the GIS community will also have an impact on the kinds of GIS that are readily adopted by business. Since 1981 census data were released, GIS have become more sophisticated and applications have become much more client-specific. Census data will be held at much more disaggregated levels by data bureaux than has ever been the case before. Clients are now able to use the information to answer wide-ranging problems and to tailor applications of such data within GIS towards their own particular problems and issues. It is very possible that in the future we will see European-wide databases which 'pool' census data from each European country (see Waters, Chapter 3, this volume). Recent debates about Euripides, the European Union initiative on pooling geographical data, highlights the potential importance of this issue for GIS. In the future GIS will therefore need to handle several types of geographic unit at varying levels of disaggregation.

Customized and tailored uses of GIS systems will both have a strong presence in the retailing environment of the 1990s. Improved methodologies, new software techniques and in particular the huge fall in the cost of computer processing power will allow the GIS industry to offer solutions that would have been out of the question even a few years ago. This improved public-information base coupled with increased availability of large, constantly updated client databases also points towards much more customized classification systems in the future which are either industry-specific of indeed specific to the individual client who has his or her own very specialized problems. This obviously has implications for the development of proprietary systems.

Many large organizations now have data on their own customers – products purchased, payment history and so on. By merging this information with census and other data, it is perfectly possible to develop classifications of customers. Customer classification will become more important as a GIS tool in the future as customer databases improve. As the information services market becomes ever more sophisticated and competitive, it is only natural that suppliers will move towards the development of niche products. Classifications intended for use within specific industry sectors are a predictable consequence. Such systems have some merits, and with products such as FINANCIAL MOSAIC from CCN (see Appendix C) and FiNPiN from Pinpoint, the GIS industry has recognized the value of this approach.

However, the overall benefits of industry-specific classifications are less than might appear at first glance. There is a supposition with industry-specific classifications that certain types of neighbourhood do not need to be kept in separate clusters. Although these different neighbourhoods may share common levels of usage of the industry's products or services, this does not mean that there will not be significant differences within clusters in terms of the sources from which people buy products, the forms of advertising that influence them and the prices they are prepared to pay. Most industry-specific classifications end up being used not as multivariate classifications but as simple one-dimensional rank ordering systems. If a ranking of areas on the basis of a single dimensional measure of opportunity is what is needed, a regression-based system would typically deliver better results.

The practical benefits to be gained from the use of a classification system depend upon the level of back-up support, for example survey links, applications software and industry acceptance. Many GIS experts hold the view that a single national classification system would be preferable to a variety of competing generic systems. One can clearly see the problems that might be caused by the proliferation of niche systems. Fortunately, the standard proprietary geodemographic discriminators are relatively cheap to use and do provide consistent coverage of the country.

9.6 GIS futures

Given the range of issues that retailers are now addressing and the more complex problems that GIS will need to consider, proprietary system development must proceed by responding very sensitively to the needs of the client. Only in this way will development remain relevant. There will be a greater role for additional support services alongside software systems and a greater need for tailored and customized solutions for particular problems.

There is now an abundance of retail space. This can only generate greater competition amongst retailers. When things get tougher, retailers will adjust their strategies accordingly. This will lead to a greater emphasis on micro-marketing tactics, such as the adjustment of a branch's offerings in terms of style of layout, product mix, etc., to provide for the geodemographic profile of the local catchment. Many of the major brands of retailing that have been built up in the last 15 years or so will begin to use geographical fine-tuning techniques to ensure that the branch is maximizing its opportunity in its own micro-market. Along with this will come the development of more niche retailing, in which retailers pick a very narrow market sector and design their offering to trade in just that sector. In both of these cases, a detailed knowledge of catchment areas, and their demographics, will be absolutely vital. In other words we should anticipate a greater use of GIS tools and database management in the future.

9.6.1 Data integration

Since about 1980 a whole industry has grown up around the use of geographic information in business applications. The magic word 'geodemographics' was deemed in the 1980s to provide the elixir of eternal success. In fact, for a decade, the market analysis industry followed the classic product growth cycle. There are now signs that this relatively new industry is reaching maturity. It is no longer looked upon as the 'Cinderella of marketing'. Many marketeers now regard geodemographic and other GIS-based analysis as an integral part of the task to maximize sales potential through accurate consumer targeting.

There are certainly gains to be made in the development of better classification systems and better software for handling geographic data. However, the major leaps forward will come as work is done to address the issues of data integration. Aspects of the data integration task are illustrated by my own consultancy experience in the area of turnover prediction and retail performance analysis.

Retail performance modelling

The use of graphical displays to front-end large GIS can produce very showy results. But it is often said that you can get away with murder on maps if you do not know what is in the 'black box' that produces them. The inner workings of the models inside the black box are mysterious to many people. On the other hand enhanced GIS software provides a unique opportunity. For the first time, GIS makes it practical for many clients to use sophisticated modelling techniques.

For example, regression and gravity models can be harnessed to provide meaningful and focused bench-mark answers for their users, and geography and demography are major components. However, in the past, relatively few organizations have benefited from their existence. For many retailers the cost of collecting appropriate data, hiring high-level expertise and developing a

217

model is beyond their means. However, as models become more easily configured in off-the-shelf business GIS software it is likely that they will be used more commonly within the industry. It is when researchers can access GIS software that can capture the power of different models, without having to do anything other than define or reference their data-sets, that they will recognize the value of this technology.

Geodemographics is not the only determinant of a successful site. Although the ability to discover what kinds of people are living around a store was a major leap forward, the market is now more demanding. Retailers want to be able to link geographic information to many other characteristics about a site. For example:

1. Micro-site – position on the high street, traffic flows, pitch quality, location of nearby shopping magnets, car parking.
2. Store details – window displays, floor-space, store appearance, design and layout.
3. Staff details – number of staff, quality and training.
4. Competition – presence of major competitors in the micro-environment and in the larger catchment area.

Although many of these facets are not directly geographical, the need is to be able to link these with geographical data to produce more realistic potential models and turnover prediction models. The data to measure some of the above determinants are now more readily available than they were ten years ago. For example, external data suppliers have collected data on the location of major retailers which can be used for competition analysis, and computerized payroll and staff-management systems now hold data within companies to address staffing characteristics. Store details are also more readily collected, especially for merchandising purposes, and there is a greater willingness to collect other information via store surveys for turnover prediction analysis.

Statistical techniques have always been available to analyse such data for turnover prediction but what has always held retailers back is the lack of data and usability within the commercial arena. Regression analysis is a useful technique for integrating data from diverse sources into a turnover prediction model but these techniques can only be more widely used when clients can use them in-house rather than to have to return to a team of consultants whenever they wish to assess a particular site. This very problem originally led SPA to write user-friendly software to accommodate regression analysis for one particular client. Originally, therefore, this was an example of a very customized approach. However, we always knew that there were a number of our clients who wanted this kind of capability and who had the right kind of data and a high enough number of branches to make turnover prediction models valid.

The product has developed into a generalized system which allows the results of regression analysis to be easily displayed and used. This highlights the more general point that customized software can be generalized to multi-user client requirements and thus become generic. Obviously it is still a very specialized project to build a regression model for a particular client and by its very nature it has to be customized. However, retailers wish to use them for similar purposes and this has led directly to the development of our own proprietary system for turnover analysis and displaying of associated results. Within this system, geographical data sit alongside other data, allowing the retailer to integrate information from a wide variety of sources. Figure 9.2 shows a screen of information from SPA's regression analysis system – a customized solution within a user-friendly software environment.

Retail analysis systems

As another example of the impact of data integration on the future of GIS one only has to look at the vast volumes of data now being generated from electronic point-of-sale (EPOS) systems. Many of these data have a geographical component as they can be associated with the locations of individuals on the one hand, but on the other can be aggregated to branch level, sales area or region, or indeed any level of geography that is appropriate to a particular business.

The key problem that this mass of data generates is one of being able to extract relevant information quickly and easily, in order to answer specific questions or to produce standard management reports. An opportunity exists to develop systems that have a geographical component which will be able to sift this vast amount of information and to produce the appropriate statistics, often in the form of charts, other graphics and maps. Usually the requirement is only for simple cross-tabulations, bar charts or performance plans. This would be easy if the information load were small. However, the challenge is for such systems to offer the facility to carve up large and complex data-sets by any dimension, and then to display the results in a form that can easily be assimilated.

A major development area for the GIS industry will be the development of systems to allow retailers to dissect their data by any of the components within it: for example, to look at sales of particular products over time and by location. Integrating data in this fashion allows inter-relationships and patterns to be highlighted and allows the user to understand the business more fully. Some of the questions which could easily be answered are:

1. How do sales of product x vary between my branches in sales region 1?
2. How did the promotion of formal shirts affect sales of casual shirts in my 'younger style' branches?
3. How do actual sales compare with potential turnover in my Northern Sales Region and does this differ within merchandise groups?
4. What were sales margins like in the comparable period last year in branches that have now been refitted?
5. Does my town classification help to differentiate my poorly performing stores from my better performing stores?

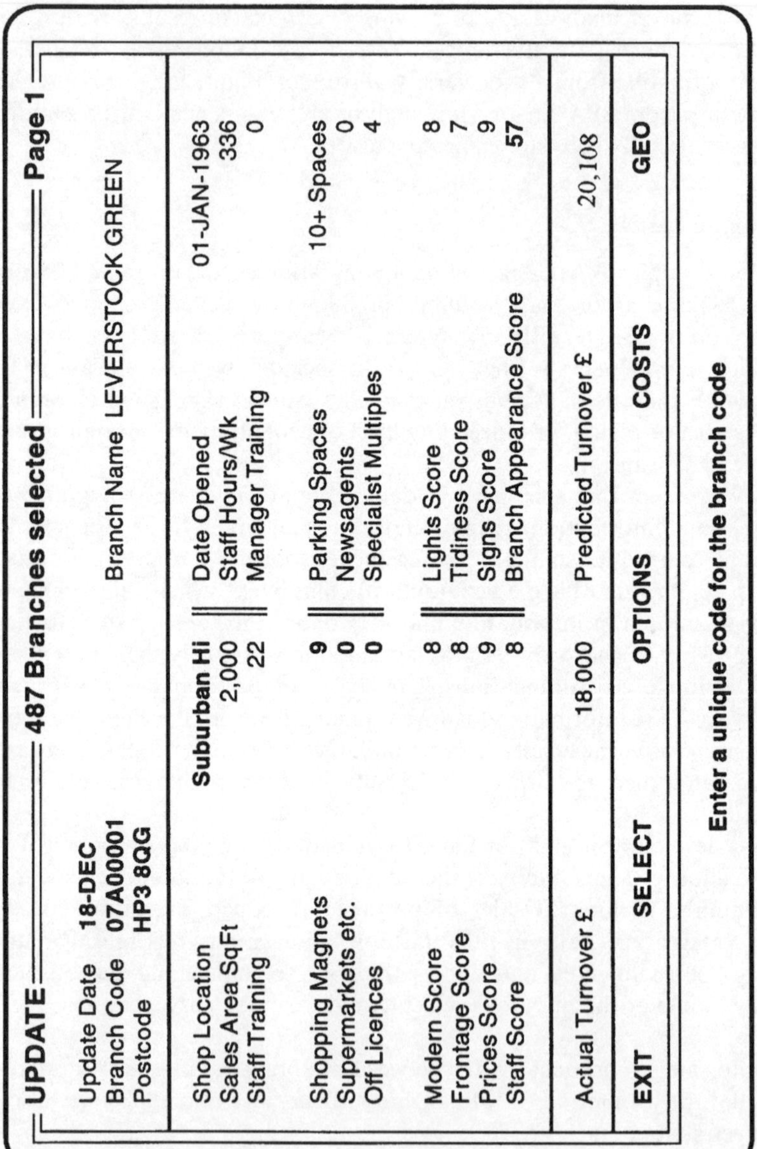

UPDATE ══════════ 487 Branches selected ══════════ Page 1		
Update Date **18-DEC**		
Branch Code **07A00001**	Branch Name **LEVERSTOCK GREEN**	
Postcode **HP3 8QG**		
Shop Location	**Suburban Hi**	
Sales Area SqFt	2,000	Date Opened 01-JAN-1963
Staff Training	22	Staff Hours/Wk 336
		Manager Training 0
Shopping Magnets	9	Parking Spaces 10+ Spaces
Supermarkets etc.	0	Newsagents 0
Off Licences	0	Specialist Multiples 4
Modern Score	8	Lights Score 8
Frontage Score	8	Tidiness Score 7
Prices Score	9	Signs Score 9
Staff Score	8	Branch Appearance Score **57**
Actual Turnover £ 18,000	Predicted Turnover £ 20,108	
EXIT	SELECT	OPTIONS	COSTS	GEO

Enter a unique code for the branch code

Fig. 9.2 A screen of information from SPA's regression analysis system

9.6.2 Software platforms and database design

The rigid structure of early GIS was often determined by the software languages that they were written in. The future is exciting for GIS because of the new ways of handling and storing data and the new software languages that make flexibility, openness and data exchange possible. It would be hard to imagine the kind of applications discussed in this chapter being developed for PC applications ten years ago, particularly with the size of data-sets that such systems are now required to handle at a respectable speed. Enhanced graphical capabilities have improved user interfaces, so much so that all good GIS should now be easy to use.

The increasingly wide adoption of the Windows computing environment is particularly relevant, given its open structure and its ability to link into other proprietary software such as spreadsheets and powerful graphics packages. Linked to powerful database engines that allow cubes of seemingly unrelated data to be analysed, software icons such as Windows will allow new, more powerful GIS applications to be developed. This is very much the way that the US market is developing, with many GIS being written in this open architecture so that data may be swapped between applications and integrated much more effectively. The 'Tactician' GIS package from Tactics (USA) is just one that demonstrates the flexibility that is offered by such software platforms. The environment allows the user to switch easily between maps and reports, and to draw on spreadsheets to integrate the data, models and functions that help retailers in their search for maximum sales potential.

Tactician connects executives, brand managers, and analysts to data using the dynamic, spatial perspectives of high-speed maps. Connectivity with many different spreadsheet packages and data formats enables the product to link easily with other user applications. The operator can click anywhere on the screen using a mouse and in an instant the flash report gives him or her a view of data underlying the map, such as customers, service centre, or retail site.

An icon editor allows the proprietary software to be customized, in order to build symbols and logos which customize the display of point data such as stores, sales representatives, or customers. Linked to the database, the icons can be sized and queried for 'on the fly' information reports. Tactician's (Tactics International: see Appendix C) high-speed TIGER Street System enables users to locate an address anywhere in the US and to display streets with their names. A key feature of future GIS will also be their connectivity to other operating systems and databases – a feature ideally suited to the Windows environment. The ability to integrate the dynamic of a map to reference many spreadsheets simultaneously, as Tactician does, will need to be a feature of proprietary systems, with complete query systems for marketing functions. Plate 9.3 depicts a screen from the Tactician product and conveys the idea of a multi-functional GIS controlled within the Windows environment.

Drivetime Inc, the SPA Group's US office in Boston, has been set up to exploit the GIS potential of Tactician, not for traditional GIS marketing applications, but for planning sales territories, rationalizing sales forces, and journey planning using drive times. Companies in the US, such as the giant greetings card company, Hallmark, have invested in this sort of GIS for the whole of the United States. The Company has over 800 sales territories covering some 43 000 five-digit zipcodes. Tactics SPA has used a computerized national highways network, supplied by Spatial Data Sciences, to create a database of drive times between zips for each of the company's sales regions. These can then be used to plan sales territories effectively.

9.6.3 Expert systems and neural technology

One area which is beginning to attract interest in the GIS industry is the development of expert systems concepts in a GIS framework. Initial experiments in the US with marketing applications of expert systems look encouraging. Expert systems simulate human reasoning to solve problems in areas characterized by inexact and ill-structured knowledge, and recent work has tried to link expert systems with spatial reasoning (see Zhu and Healey 1992; Lein 1992). By constructing 'influence diagrams', problems such as the suitability of a site for a retail store location are broken down into their influencing factors until the data required to solve the problem have been identified (see also Openshaw, Chapter 7, this volume). These data can be acquired through demographic information suppliers, the client's own data sources, survey analysis or databases of competitor activity.

A meaningful set of values must be defined for each of the numerical decision factors in the influence diagram, and the relationships among the factors must then be captured in 'rule tables'. These are directly translated into the knowledge base of a geographical expert system. The knowledge base can then be used to analyse a geographic area and to produce a map representing the problem at hand. The results of this analysis can be explained for any selected location on a derived map and the explanations provided by the system can be used to validate the model of the problem. After an initial knowledge base has been constructed, its knowledge can be iteratively refined until the user is satisfied that it adequately represents the best available expertise on the problem.

Along similar lines, the application of neural network technology is now beginning to emerge in marketing circles, as a tool for data analysis (see Openshaw, Chapter 7, this volume). Recent discussions have explored the use of neural nets in customer segmentation, retail modelling and sales analysis. One of the main advantages of neural technology is that neural nets provide a way of modelling patterns in data without having to make any assumptions about the underlying reasons for the patterns. With the development of neural net software, new proprietary systems may yet emerge based on these new techniques. However, business applications of neural technology are still relatively scarce, and to this author's knowledge have never been documented, and so their utility in the business arena has yet to be comprehensively demonstrated. It remains to be seen whether such 'expert concepts' will be translated into commercially available systems on a wide scale, but there is certainly potential for improving spatial modelling in GIS contexts.

9.6.4 Global GIS

With the emergence of the Single Market, pan-European, and indeed global, applications of GIS software and databases are now a reality. The future should see a greater degree of flexibility in software and databases, for addressing the pan-European or worldwide structure of many organizations and for centralizing control and providing easy access to data via a single system. There is vast potential for the application of GIS techniques for marketing increasingly international brands outside their original home markets.

The trend towards global databases is starting to be recognized in the provision of data which seek to give some consistency to the many varied data sources, in terms of quality and quantity, which exist throughout Europe (see Waters, Chapter 3, this volume). CCN, for example, have recently produced a version of their proprietary MOSAIC classification systems called Euro-MOSAIC, allowing consistent classification of 310 million European consumers on the basis of the types of neighbourhood in which they live. EuroMOSAIC groups each country's neighbourhoods into a set of ten major classifications, each of which contains a consistent set of demographic, social and housing characteristics. Groups such as 'Elite suburbs, 'High rise social housing', and 'Vacation/retirement' areas can now be identified on a European-wide basis from a centralized control point – notwithstanding Birkin's doubts (Chapter 6, this volume) about the detail of their discriminatory powers. EuroMOSAIC does provide an immediate feel for where to target resources in an unfamiliar market and allows the degree of similarity in brand profiles across each national market to be examined.

Many countries now have classifications of their own markets, for marketeers who want more detail for specific countries. As well as the pan-European EuroMOSAIC, MOSAIC is also available at a more detailed level for several countries (see also Waters, Chapter 3, this volume, for an account of public-sector data sources in many of these countries):

1. Netherlands. Dutch MOSAIC is built for the Netherlands's 390 000 six-digit postcodes, from several sources. These include: the PTT central postal directory, which gives housing type and tenure information; Wehkamp's (mail order company) useful statistics on mail order responsiveness, age and credit worthiness; and market surveys, which provide extra demographic data.
2. Belgium. Belgian MOSAIC classifies over 120 000 streets into a series of lifestyle types, built using statistics from the 1981 census. The information is supplemented by more recent information on car ownership, family and age structures.

3. Sweden. Swedish MOSAIC classifies the households in each of Sweden's 5557 postcodes into 29 lifestyle types. Information comes from the 1986 census and covers a range of demographic variables.
4. Republic of Ireland. Irish MOSAIC classifies every one of Ireland's 80 000 streets into a specific type, describing the lifestyles of the households. Data come from a variety of sources including the 1986 census, the 1990 Electoral Roll, credit activity sources and a shareholders database.
5. Northern Ireland. Northern Irish MOSAIC classifies households according to each of Northern Ireland's 41 000 postcodes. Data come from 1981 census data, 1986 Northern Ireland Housing Executive Survey data, 1990 CCN directors' and shareholders' data, 1990 Postal Address File and 1990 planning applications data.
6. Spain. Spanish MOSAIC classifies each household in each of Spain's 35 000 Electoral Sections into lifestyle types. The system uses data from the 1989 Electoral Census, which provides detail on the adult population's socio-economic characteristics.
.7. Germany. German MOSAIC is constructed using individual house surveys, lifestyle surveys and government statistical data.
8. Australia. Australian MOSAIC has been developed at both Census Collection District and postcode level, in order to enable targeting at two levels of geography. Data come from the 1986 National Census.

All of these classifications can be interrogated by CCN's proprietary ENVIRON geographic database manager, in just the same way that the 1991 census can be analysed in Great Britain. The full range of GIS applications is also available, such as mapping, customer profiling, catchment analysis and direct marketing. This makes global GIS analysis tools a reality. GIS software is now beginning to recognize the need to develop in a platform that is multi-national, with the ability to analyse and interrogate data from varying sources and nationalities in as consistent a way as possible. The Tactician software has a 'global' component which can manage different map projections, for example, latitude/longitude, Albers, national grids, and other units of geography (such as US zipcodes and UK postal codes), while retaining global capacity when moving between countries.

9.7 Conclusion

Recent years have seen significant developments in the tools that we, in the market-analysis industry, have available to answer the wide-ranging questions that the marketplace demands.

There is still a very important role for proprietary GIS to play within the business environment. However, the days of 'shrink wrapped' GIS products are likely to be numbered. Proprietary systems offer a very good toolkit for addressing many of the problems that retailers face in today's competitive environment. However, the nature of the problems which individual retailers face today is so varied that proprietary GIS cannot hope to provide all of the tools. This points towards increased guidance from the GIS industry to help clients use the relevant range of tools at their disposal in order to meet the challenges that they face. In some cases this will involve the development of very specialized and customized geographical systems. On the other hand it often means helping the client to use proprietary systems more effectively and to tailor the databases that such systems rely upon to be much more relevant to each client (see Clarke and Clarke, Chapter 10, this volume). Guidance in helping clients to find new ways of using existing systems will undoubtedly be an important growth area for the GIS industry.

On the part of potential users of GIS, whether proprietary or customized, the failure fully to exploit GIS is now, unlike the early 1980s, increasingly a non-technical issue. Rather, it is the information literacy of the managers and the information culture of their organizations that are the key determinants. Academic research into GIS-related issues must be set within the context of the ever-changing problems that retailers are facing. It is often difficult for research programmes to track these ever-changing issues. Academia has to try to respond to these issues and to work closely with the GIS industry to translate ideas into commercially viable products or projects (see Clarke *et al.*, Chapter 12, this volume). Only if new ideas can be used by retailers 'on the ground' quickly, efficiently and in a user-friendly framework will the wider dissemination of academic research into the business community become possible.

Above all, the development and future of GIS, whether proprietary or customized, must at their heart concentrate on the integrity and validity of the data that are used to answer business problems. This feeds through into calling for greater industry cooperation between GIS suppliers, consultants, and bodies such as census authorities and national postal organisations. For example, clients would rather not be confused by the range of geographic building blocks at which census data are supplied, and likewise the GIS industry would benefit from closer cooperation as one body with government and other data suppliers to our industry. The population surface modelling technique described by Martin (Chapter 4, this volume) may offer a robust solution to some of these problems.

GIS for today's fast-moving retail environment needs to be flexible enough to answer a range of analytical issues. Given the range of issues that now need to be addressed there is a danger of GIS software houses developing systems that are further from, not closer to, the end-user client's needs. Developments in commercial companies as well as within academia must not lose sight of the problems that retailers are actually having to address on the ground. All good proprietary systems have developed directly as a result of listening to clients and looking at ways to address particular problems (see Maguire, Chapter 8, this volume). Often many retailers have the same kind of problem but within different circumstances. This development route leads directly to ideas for new products which address the needs of the market precisely.

The rest of the decade will also focus greater attention on the micro aspects of marketing and will rely on detailed knowledge of each local catchment area in order to formulate a policy specific to particular catchment areas. GIS proprietary software is now sufficiently advanced to be able to provide this information at the client's fingertips and to unlock the power that these systems have across many market sectors.

10

The development and benefits of customized spatial decision support systems

Graham Clarke and Martin Clarke

10.1 Introduction

10.1.1 Context for this chapter

For retailers of goods and services distribution is the crucial component of their businesses. Their products are delivered to the consumer market through a set of distinct and discrete units: shops and supermarkets, bank branches, motor dealerships, fast-food restaurants or even schools and hospitals. Each of these units comprises part of a supply network with its own distinct geography. Investment in the distribution channel is often large as significant competitive advantage in terms of brand, format and location can generate significant profits. During the 1970s and 1980s this investment was largely associated with network expansion as the main multiple retailers, supermarket groups and financial service organizations battled it out for increasing market shares (Kay 1987). As the recessionary climate of the 1990s does not necessarily encourage such expansionary strategies, then the battle for market share (and profitability) shifts to the search for greater returns from the existing store network. For this reason, it is essential for retailers to understand the performance of their network and of individual outlets within that network: only then can they assess the potential for changes to that network and forecast and monitor the threats from competitor market strategies. To achieve this capability, which is at the heart of most corporate strategies, requires information and intelligence on what has been, what is, what if and what should be.

Geographical information systems (GIS) have been put forward as offering the solution to such information requirements (see also Maguire, Chapter 8, this volume, and Cresswell, Chapter 9, this volume). The aims of this chapter are to evaluate the existing use of GIS in business and to argue for the development of spatial decision support systems that deliver 'intelligence' that can be used for competitive advantage. In Section 10.2 we suggest reasons why proprietary GIS is failing to make significant inroads into the business community. In Section 10.3 we put forward our ideas concerning 'intelligent GIS' and illustrate the applied importance of adding greater analytical power through a number of key business questions. We suggest that the solution to such needs lies not within conventional GIS but in the form of 'spatial decision support systems' (SDSS) based around the business process. In Section 10.4 we assess the benefits of GIS and SDSS through a number of short case studies. Some concluding comments are offered in Section 10.5.

10.1.2 Context for the authors

Given that some of the comments made in this chapter will be somewhat at odds with some of the sentiments expressed in other chapters of this book, it is perhaps worth setting the context of the authors' backgrounds and experiences that have resulted in the views expressed in the rest of this chapter. While our main academic interests have largely been in the development and application of mathematical models of urban and regional systems, we have both been extensively involved since the mid-1980s in the development of GMAP Ltd, a company owned by the University of Leeds, UK. GMAP develops customized spatial decision support software for a wide variety of blue-chip clients in Europe and North America. Examples of this work are reported in Birkin *et al.* (1995). During this period, our discussions with senior executives of many large corporations have led us to try to understand the range of problems that businesses face, particularly with regard to the geographical dimension of their organizational structure.

The success of GMAP, which in 1994 employed 70 staff and has an annual revenue of £3.5 million, has convinced us of a number of things. First, that intelligence about the geographical dimension of a business is essential both to understanding its performance and, crucially, to planning for its future. Secondly, the application of GIS or SDSS or any other such approach does not take place within a vacuum but within a set of well (or poorly!) defined business processes. Failure to understand this context is a recipe for disaster. Our strongly held views come from experience. They cannot always be supported by evidence that is published in the usual sources (journals, books, etc.) but they are no less valid for this. Indeed, it was and is the very challenge of introducing geographical analysis methods to the business community, very much a non-traditional academic endeavour, that has underpinned the growth of GMAP.

10.2 Why is GIS failing the business community?

GIS applications in the UK are highly segmented (see Maguire *et al.* 1991), and it is perhaps unfortunate that such a diverse range of applications gets labelled under a single generic name. From the wide range of applications in the literature it would be fair to say that the most impressive (in terms of problem solving) lie in the fields of digital mapping, environmental monitoring, and within network analysis. GIS applications in the economic–social research areas still lag far behind their environmental counterparts. There are signs that this is changing in the US, especially through the application areas outlined in Castle (1993a) and Sherwood (Chapter 11, this volume). However, from our extensive experiences, awareness of GIS at an executive level in the UK is virtually non-existent. For some, the initial seduction of seeing 'maps' of their business and markets is quickly replaced by a scepticism towards yet another technology-driven area. Where GIS has been introduced it has tended to be seen as a technical operation tool rather than a vehicle for strategic planning. Despite the hype from vendors (and some more committed academics), we believe that proprietary GIS will not fulfil its potential as a business planning tool unless it fundamentally changes and pays attention to business processes. GIS, and in broad terms the GIS industry, is failing the business community because, for various reasons, it has become a commodity which is increasingly traded on price rather than benefits. Although GIS vendors will be quick to respond that they are making substantial investments in new analytical functions within GIS (see Maguire, Chapter 8, this volume), there is a strong convergence of technical functionality which makes the choice of GIS move closer to one based on price (and perhaps visualization quality). The GIS vendors have made matters worse by emphasizing this technology rather than focusing on business processes. The end result is a great deal of user confusion and ignorance. To illustrate this divergence between what business managers actually require and what traditional GIS provides, we can examine three major issues facing UK businesses in the retail sector. These include:

1. In the supermarket sector the growth of continental European companies such as 'Aldi' and 'Netto' that are extremely price competitive (see Wrigley 1993). They have done much to erode the high margins that the British supermarket giants came to enjoy throughout the 1980s. Coupled with this development is the recent UK opening of the first American 'Costco' retail discount warehouse club in the UK at Thurrock, offering significant savings to members who are prepared to buy in bulk. In the transformation that is taking place in the supermarket sector a range of important questions is being asked about how to respond to the new challenges. We shall argue later that to add value to this process GIS advocates must understand the context in which it might be used rather than assuming it has something to offer.

2. In the automobile industry, cars are retailed through a selective distribution system in which the manufacturers appoint franchises with a specified and exclusive geographical territory. These franchise holders are expected in return to sell only that manufacturer's product in their dealership. This system is actually in contravention of the European Community's Treaty of Rome (as it is anti-competitive), but the motor industry was block-

exempted from the Treaty for ten years from 1985. In 1995 the Commissioners of the European Union will review block exemption. The consensus view is that there will be some relaxation of the rules so that greater competition will be encouraged. What the consequences will be for new-car retailing, and the geography of manufacturer performance, are questions that future spatial analysis must address.

3. In the financial services industry there has been a major change since the 1985 Financial Services Act effectively deregulated the industry. Greater competitive pressures, coupled with developments in information technology and communications, have changed the way in which products and services are delivered to the market. ATM networks have appeared, branchless banking has established a permanent market presence, and branches are being closed at an increasing rate. All of these produce major geographical interests which must be addressed within our GIS environments.

In these three examples (and there are many more!) there is a common thread – that of understanding the structure, process and dynamics of geographical markets. In all cases, however, it is important to start from an understanding of the business context. Once this has been achieved, then the necessary tools for analysis can be assembled. These tools may well include a traditional GIS capability but the point to emphasize is that GIS is not necessarily the starting point. Part of the problem is the fact that GIS development has been technology-led. The spatial analysis functions now synonymous with GIS (polygon overlay, buffering, etc.) have not been the results of discussions with end-users in the business world. All of us working with spatial data and their analysis have accepted the vendor-driven functionality within GIS, and any other forms of spatial analysis are constantly rebuked as being non-GIS. Thus, businesses have been forced to shoe-horn their geographical problems into the black-box functionality of proprietary GIS.

There are a number of classic examples in the literature. In retail analysis, for example, a possible new site would be identified and the GIS would be used to predict likely revenues for the new store. Once population information is input, the user can buffer travel times around the new store and then calculate the population within each time band using the standard overlay procedure. This is illustrated well by Beaumont (1991), Howe (1991), Elliott (1991) and Reynolds (1991) amongst others, and is shown in Figure 10.1. The real difficulty is then how to transform population total figures into branch sales. The most likely method is 'fair share' (Beaumont 1991), because of the great difficulty of dealing with the competition using these sorts of methods. Hence, if there are three other competing stores in the buffered catchment area of the new store then the new store may be expected to obtain 25 per cent of the revenue generated in that catchment area. This simple fair-share allocation could be weighted by store size or by retail brand to increase realism. However, this sort of procedure simply cannot cope with the complex set of real interactions between residential areas and retail locations which are distorted by

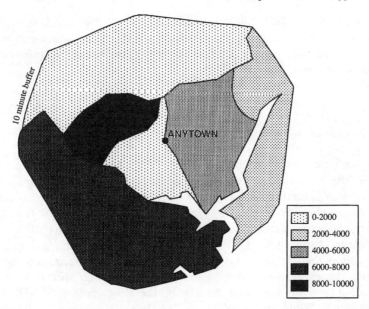

10 minute buffer

ANYTOWN

	0-2000
	2000-4000
	4000-6000
	6000-8000
	8000-10000

Fig. 10.1 A typical overlay and buffer analysis for retail catchment assessment

intervening opportunities. As Elliott (1991) acknowledges, the presence of competing centres will restrict the catchment boundary of a new store in certain directions. Her response is to 'override the drive time where it seems appropriate' (1991, p. 171). The authors of these types of studies (mainly retailers themselves) claim such analysis *is* useful and we must take this on board. However, the limitations of such catchment area studies will always produce very partial solutions because of the difficulties in coping with the allocation of demand to individual stores (both within the client's chain and that of the competition). More importantly, as we have demonstrated, business needs are far more complex and demand analytical functionality which can cope with that complexity.

In the following sections we argue that these levels of complexity can only be addressed through more sophisticated spatial models that are purpose built to handle spatial interactions. We should note that a number of GIS designers have now incorporated such routines into their proprietary systems (see Maguire, Chapter 8, this volume). One might expect us, therefore, to welcome this move. However, if such spatial models are to be used for serious business applications from within proprietary packages it is crucial that they are flexible enough to be parameterized and calibrated for each new application area. The difficulty with the spatial-interaction models in ARC/INFO, for example, is that they are aggregate models driven by a single parameter. Models required for applied research (where the expectation is to reproduce existing and future turnover within 10 per cent) will usually have to be highly disaggregated to capture the complexity of existing consumer behaviour. Having single models available as off-the-shelf solutions may actually be extremely dangerous and lead to spurious applications which will do the discipline of geography little service. The solution to generic versus customized analytical functions remains an ongoing research task.

10.3 Towards 'intelligent GIS'

The argument of the preceeding sections is that business organizations require more in the field of analysis than current proprietary GIS can provide. Whether such analytical power can be built into proprietary systems remains one of debate (Fischer and Nijkamp 1993). One solution is to look towards the concepts of 'spatial decision support systems' (SDSS), or what Birkin *et al.* (1995) term 'intelligent GIS'. This involves integrating existing GIS functionalities (particularly those relating to storage, visualization, aggregation and disaggregation) with spatial models through direct or indirect coupling. The definitions and history of the development of SDSS from management schools are supplied by Densham (1991) and Densham and Rushton (1988). They reiterate the fact that spatial problems are generically complex and are usually ill-defined or semi-structured. For these reasons we need systems which can handle a large variety of different business problems in a user-friendly environment (i.e. no visible computer commands other than a start procedure). Typically GIS is not enough here because of the lack of such suitable analytical functions. What is required is the ability to link spatial models with the strengths of GIS in storage, retrieval and graphics. The philosophy of spatial modelling recognizes that there are three main components of a market system: demand for a product which is spatially variable across population groups (differentiated by income and/or lifestyles – see Batey and Brown, Chapter 5, this volume; Birkin, Chapter 6, this volume); supply of a product through discrete locations such as shops or supermarkets; and *interactions* between demand and supply. The aim of spatial modelling is first to reproduce or estimate (if unknown) the patterns and magnitude of consumer interaction between demand locations (households or groups of households in census tracts or postal areas) and supply points. In the car market this interaction set needs to be disaggregated by brand and market segment, which can produce levels of great complexity. Consider, for example, the study area of Seattle in the US (see Figure 10.2). This has 500 census tracts (demand zones), 100 retail car outlets (supply zones), 10 main manufacturers, 5 household types (population by income) and 5 main market segments (family car, sports car etc.). This gives $500 \times 100 \times 10 \times 5 \times 5$, or 12.5 million, potential interactions. With growing computer power spatial models can handle this level of complexity with ease. The development of spatial interaction models has a rich history in the geographical literature. Rather than restate the evolution process behind the development of this type of modelling, we refer the reader to Wilson (1974), Clarke and Wilson (1985) and

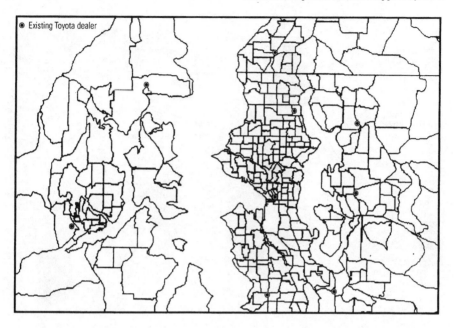

Fig. 10.2   Seattle, USA: locations of Toyota car dealerships

Birkin *et al.* (1995). Such calculations of interaction patterns provide the starting point for a wide range of support routines tailored to real business requirements. The central question to be answered in the remaining part of this section is: what can intelligent GIS do that current GIS cannot? We illustrate this by examining a number of key business issues.

10.3.1   How well are we meeting market needs?

For some organizations data on market performance are available industry-wide and hence market shares can be calculated quite easily. Plate 10.1 shows such a situation for the car market. Here, sales have simply been divided by the total market within each postal district to give small area market-penetration values. If such data are available then proprietary GIS will be able to calculate and map market penetrations. However, many organizations are not so rich in customer information. For these, market-penetration values have to be estimated using spatial models. It is unlikely that existing overlay methods can adequately complete this sort of estimation because of the difficulty of trying to calculate the revenues of all competitors. Figure 10.3 shows an example of market penetration for a leading grocery retailer in North-East London calculated by spatial models. By overlaying store locations it is possible to see the strong correlations between market penetration and outlet

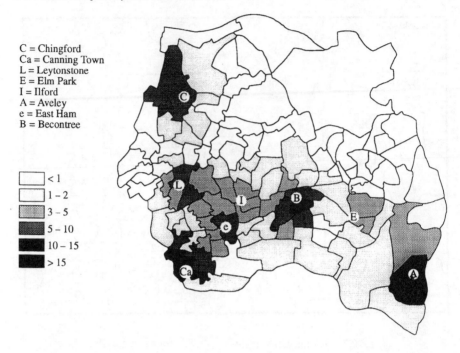

C = Chingford
Ca = Canning Town
L = Leytonstone
E = Elm Park
I = Ilford
A = Aveley
e = East Ham
B = Becontree

< 1
1 – 2
3 – 5
5 – 10
10 – 15
> 15

Fig. 10.3 Market penetration in North-East London, for a leading UK grocery retailer

location. Similarly, such maps help to identify market gaps which may be worth investigating in terms of potential new locations (or more targeted marketing/advertising).

10.3.2 How can we help understand the efficiency/effectiveness of the current network?

Although the traditional emphasis in store location research is the estimation of new store turnovers, there is increasing interest in improving the turnovers of existing stores. Indeed, some retailers (notably in the financial service market) are increasingly looking to rationalize their networks. Hence, a key question to be resolved is: how well do existing outlets perform? As mentioned earlier, business organizations are looking for much greater productivity and efficiency from existing outlets rather than basing growth on new site investments. Traditional measures for measuring effectiveness have been gross sales or, more reasonably, year-on-year sales performance. However, this ignores local market trends (i.e. the real potential for the store, given demand) and the behaviour of the competition. What is clearly required is a more objective measure of potential at a given location. This can be produced by modelling predicted sales versus actual sales. In most circumstances one would expect that such predicted sales would be within 10 per cent of actual sales. However, there will always be exceptional circumstances where a store is over- or under-performing. Comparing predicted with actual sales may thus help to give a more objective picture of store performance. Again proprietary GIS would help give the size of a potential catchment area but would be unlikely to allocate demand to individual stores with any great precision.

10.3.3 How can we help predict what might happen if the store network changes?

Understanding the current market conditions is only the first step towards an effective SDSS. There are a whole set of 'what if' questions which spatial models are ideally designed to address. Despite the overall trend towards consolidation of store networks a number of organizations are still committed to expansion programmes within particular market niches. For example, the UK discount retail chain 'Kwiksave' announced in November 1993 that it intends to open the equivalent of one new store per week for the next two years. Similarly, the fast-food retailers 'Taco Bell' in the US have a plan to open 3000 new outlets across America over the mid- to late 1990s. There is also an increasing international dimension to retail store expansion. As major home markets become increasingly saturated so retailers are looking to expand internationally. Wrigley (1993) provides examples of this type of 'spatial switching of capital' in the grocery sector. For all these types of groups (and their competitors who need to monitor such changes on their own corporate performance levels) the ability to forecast or predict likely revenues is crucial (we shall return to this in Section 10.3.5). The model should be able to predict not only turnover levels (by segment) but also a range of performance indicators related to market shares and profitability (Birkin *et al.* 1995; Birkin 1994).

Not all organizations will be interested in spatial modelling for assessing the impacts of new store openings. Many retail businesses are looking to rationalize existing branch networks. The banking industry is a good case in point. Many of the UK banking organizations now feel they have too many branches on Britain's high streets and are looking to make significant cost reductions through branch closures (especially as ATMs and self-service banking become more common). Once again it is important to make sure that the right branches are closed and the potential damage to market shares is minimized. A third use of 'what if' capabilities is to examine the impacts of refurbishments. This may involve the realignment of product ranges to match more closely the nature of local demand. In banking, for example, it may be clear that a particular branch is performing poorly in mortgage sales given the high rate of new house construction in an area and the high rate of new household formulation (and hence persons looking for a first-time mortgage). The bank may be able to respond by increasing the support services for mortgages and shifting staff away from less productive areas. Again it is possible to insert such changes into the model and to recalculate likely revenues and effects upon the competition.

10.3.4 How can we determine optimal business strategies for a given region?

Whatever the merits of overlay and buffering within proprietary GIS it is clear that they are wholly inadequate for handling optimization tasks. For a particular region or market a retailer is often interested in what the optimal level of representation should be. Once again, we have to look towards spatial models within SDSS, where there is a long history of research into optimization (Wilson *et al.* 1981). With such models it is possible to identify a set of constraints to the optimization procedure which may be varied in subsequent model runs. In the retail world the search for optimal locations may be constrained by the existing network of outlets, by the cost of investment, by certain viability levels or by the desire to maximize market share rather than revenues *per se*. We shall give an example of the type of optimization problem which is particularly relevant in the business sector in Section 10.4.2.

10.4 Appraising spatial investment decisions in terms of costs and benefits?

In the GIS literature to date there have been few attempts to undertake a traditional cost–benefit assessment of the use of GIS within organizations. The exceptions to this seem to be in the local government environments. Both Wood (1990) and Peutherer (1988) have put forward some perceived benefits of using ARC/INFO in Tacoma (US) and Strathclyde (UK) respectively. Combining these thoughts gives the following checklist of perceived advantages:

> Rapid and coherent access to data
> Decreased manual effort
> Decreased duplication of effort
> Improved reporting capabilities
> Dynamic updating of information
> Improved accuracy and resolution
> Interchangeability of data between users
> Integration and monitoring of data
> Improved productivity.

These benefits clearly relate to improvements in the efficiency of data storage and handling and the development of corporate sharing of information and resources. It is very difficult to quantify the financial savings associated with these sorts of benefits. That said, Buxton (1989), for Cardiff City Council, estimated a saving of £165 000 per annum through the introduction of GIS for automated map production alone. Similarly, Wellar (1993) estimated the costs of implementing GIS in an undisclosed local authority at £580 000 over a six-year period, whilst the benefits in terms of person–year savings came out at £1.2 million. For business users in particular it is essential to be able to estimate some kind of substantial financial benefits to be derived from *spatial analysis* if we are to convince them to make significant investments, since the production of maps *per se* is not an activity they would normally engage in Wellar (1993) offers the following logic test: 'If, after study, the evidence suggests that a GIS means better data – better information – better decision processes – better decision outcomes – better bottom lines, then, buying or expanding a GIS makes business sense and deserves favourable consideration' (1993, p. 17).

In the following sections we give four case-study examples of the sorts of benefits users can enjoy if they adopt GIS and spatial modelling methods, particularly emphasizing the 'bottom line'.

10.4.1 A home electronics rental company

The first case study that we wish to describe concerns a major British high-street retailer involved in the rental of home electronics products. The organization has close to 1000 outlets in the UK renting a range of electronic products to consumers. The market for their products has declined in recent years but the organization has been very profitable and has attempted to maximize the profitability of its asset base within this declining market. In building a spatial decision support system (SDSS) the organization was interested in quantifying its performance in the market and looking for opportunities to rationalize its network of branches in certain regions but taking opportunities to open new stores in areas where opportunities arose for the company to generate profit. Two examples of the use of the resultant SDSS exemplify the benefits that the system brings to the client organization.

The first example relates to the analysis of a new store opportunity in a medium-sized town in South Wales. Market analysis using the information system indicated that there was a potential opportunity for a store development in the town of Newport. The organization's market share in the postal sectors surrounding Newport was low and the level of competition in the town centre was reasonably weak. On this basis we used the system to assess the impact of opening a new store offering a full range of products in Newport. Estimates of the cost of acquiring a store and undertaking the appropriate refurbishment and signage, along with the annual running costs of the store including rent, staff costs and so on, was made. The spatial model within the SDSS generated an estimated revenue for the store in Newport over the next five years. On the basis of the analysis and the model predictions the store would generate a profit to the organization in the time scale analysed. However, when the impact on the organization's other stores in the region was taken into account it was demonstrated that much of the revenues attributed to the new store were cannibalized from these existing stores. This is a crucial output from spatial interaction models. Not only are they more accurate at

Table 10.1 Different predictions of customer retention following store closure programme

	Internal forecast	Actual	GMAP forecast
% retention	82	63	62

predicting turnovers than traditional GIS methods; they are also able to estimate where trade for a new store might originate. Most retail sectors have a finite demand and hence predicting potential new-store turnovers should be a zero-sum game. A new site will cause major *deflections* of trade from existing sites and retailers will want to ensure such deflections originate from the competition not from their own existing branch network. The model results were entered into the standard financial viability estimation procedure associated with any potential new site. This analysis indicated that the overall impacts of opening the new store, taking into account the costs of doing so, would in fact be negative. The organization therefore did not go ahead with the proposed development. This demonstrates that a modelling approach that takes into account the systematic effects of a store opening or closure must do so at the regional level, not just at the individual centre level.

The second example concerns a programme of store closure that the organization proposed carrying out across the country. Unfortunately the closure programme was implemented before the decision support tool was fully developed. Table 10.1 compares the organization's predictions of the impact against the decision support tool's model predictions and the actual outcome. These impacts are measured in terms of the percentage retention of customers of the branches that were actually closed. Effectively these customers would have had to transfer their accounts from a closed branch to a retained branch. As can be seen from this chart, the organization's internal estimate of the level of customer retention was wildly optimistic. However, the decision support model predicted the retention level very accurately. If the model had been available at the time of the policy decision, a different configuration of branches might have been retained and fewer customers lost. Again, this demonstrates the vital contribution that spatial modelling can make to strategic planning and investment appraisal. These benefits can be measured explicitly and used to justify the investment in generating the appropriate tools.

10.4.2 The automobile industry

The automobile industry is a sector that has long recognized the importance of geographical planning and analysis. All the main auto manufacturers distribute their products to the market via a network of franchise dealers. As mentioned in Section 10.2, these dealers are independent businesses but are allocated an exclusive geographical territory to which the manufacturer agrees not to assign any other dealer, subject to the existing dealer meeting certain performance criteria. Clearly most manufacturers aim to maximize their market share and profitability in the market. From analysis of the voluminous amounts of registration data it is clear that there is a very strong relationship between market share and dealer location. In other words the more dealers the manufacturer appoints, the greater the likely market share. However, this is traded off against the fact that as market share increases there are diminishing returns and the sales of each dealer reduce, thus affecting individual dealer profitability and, possibly, the scope for retail price discounting. As a consequence, manufacturers are trying to find a balance between maximizing market share whilst at the same time ensuring that each individual dealership is a profitable business in its own right.

Achieving this balance requires a thorough understanding of existing market performance and the ability to examine alternative scenarios through an intelligent GIS approach. Given the arguments of Section 10.2, we would not expect GIS to be able to deal with all of these business objectives and there is little published evidence of the take-up of GIS in the motor industry. Clifford (1993) provides an important exception. He shows how Vauxhall has invested in a GIS in order to improve its understanding of local performance. This is achieved by mapping sales in relation to dealers and their assigned sales territories. Figure 10.4 shows a good illustration of this. As in the retail examples of Section 10.2, sales can be overlaid with demographic information to try and understand the relationship between the sales of a particular car model and population types. The difficulty with this packaged solution is again the lack of any real analysis of the workings of the car market and the obsession with relating sales to drive time bands rather than with any explicit treatment of spatial interactions and the complicated influences of competitor location. For these reasons, and other reasons such as ease of use (Clifford 1993), such proprietary GIS solutions are not likely to be the most successful. Indeed, it is our experience that a growing number of senior managers within the car industry are now looking to decision support systems for a more complete solution.

The particular case study that we will use to illustrate the benefits of this type of approach is based on the analysis of the Toyota network in Seattle, Washington State, US. The specific question that we wish to address is: are there possibilities of extending the Toyota network in Seattle in such a way as to generate incremental sales and market share without detrimentally affecting the viability of existing Toyota dealers, and at the same time of creating successful new dealerships in their own right? These sorts of question demand a new type of analytical solution known as optimization models, which although well known to quantitative geographers are generally not available in proprietary GIS. The methodology involves embedding a spatial-interaction model within an optimization routine so that the model will systematically search for solutions which satisfy constraints input by the analyst. In the Toyota example the objective function was the maximization of market share subjected to the following constraints:

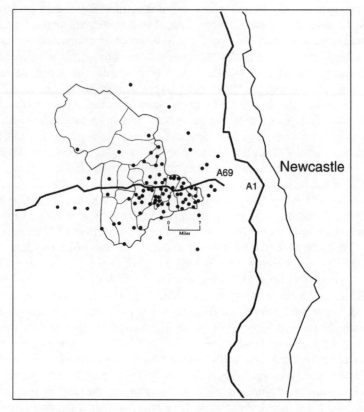

Fig. 10.4 Traditional method of 'customer spotting' using overlay operation of standard GIS

1. Any new dealer identified must have a minimum annual sales of X.
2. Any new dealer identified must be a minimum of Y minutes away from every existing Toyota dealer.
3. No potential consumer must be more than Z minutes away from a Toyota dealer.

The heuristic thus developed solved this problem and also accounted for exclusion of certain areas of the region where a location would not have been feasible – for example a park or a census tract zoned for residential use. The illustration examines two different problems. In the first we wish to identify the optimum location of two new dealers in the Seattle area which will have a minimum sales throughput of 450 cars per year and be more than 13 minutes' drive time away from an existing Toyota dealer. In this case the solution identifies two dealers which will generate incremental sales of almost 750 Toyota vehicles in the Seattle region (see Figure 10.5). The second example relaxes the drive-time criterion to identify two new dealers with a minimum of

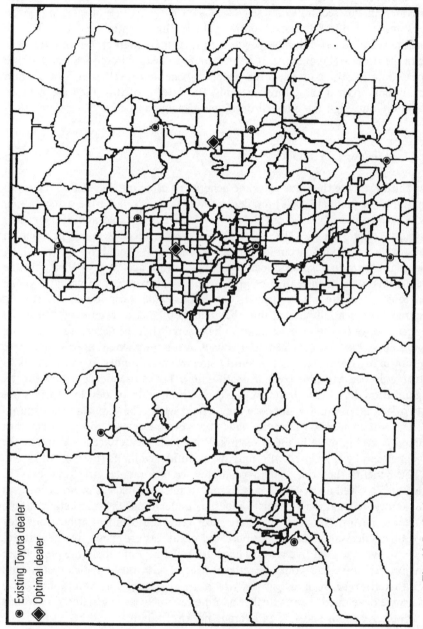

Fig. 10.5 Optimal location of two new dealerships (at least 13 minutes' drive time between dealers)

⊙ Existing Toyota dealer
◆ Optimal dealer

600 sales but at least 10 minutes drive time from an existing dealer. In this case a solution was identified but the locations were markedly different from the first example (see Figure 10.6). However, almost 1000 incremental sales were generated in this second example. Based on 1000 incremental sales and given an estimated profitability of $1500 per vehicle, the potential benefit to Toyota is in the region of $1.5 million per year in this single market. Given that there are in the order of 80 markets of a similar size to Seattle in the US, the exercise repeated across the country could generate benefits of well over $100 million per annum. Even if the figure was only 10 per cent of this, then the benefits would far outweigh the costs of developing such a system.

10.4.3 A major DIY retailer

The third example that illustrates the benefits of using intelligent GIS relates to a major do-it-yourself (DIY) retailer in the UK. This retailer has a substantial network of outlets throughout the country and has a stated aim of increasing its market share in the DIY market. However, it faced a major problem in being able to predict accurately the revenues that would be generated by opening a store in a specific location. The historic accuracy of their revenue predictions were in the range of plus or minus 30 per cent of actual revenue generated at a site when opened. Given this wide variation in forecasting accuracy, it essentially meant that they had to predict a revenue of at least 30 per cent above the break even point to be assured that the new store at that site would be viable. As most people involved in retailing would appreciate, there are not many opportunities to identify sites where revenues are so great as to return a 30 per cent-plus level of profitability. The consequence of this for the organization has been that they have not been able to develop their store network at anything like the pace they would wish. The importance of spatial models is their ability to predict new store revenues closer to the 10 per cent accuracy level in most business sectors. With such greater accuracy, new store revenues can be calculated with narrower confidence limits (margins of error).

The benefits that this system generates for the retailer are quite easy to calculate. As well as having timely access to relevant market information and the ability to manipulate this to better understand the markets and the retailer's performance in it, there are very important implications for investment decisions. Let us suppose the organization opens five new stores per year over the next five years. These stores would not satisfy the plus 30 per cent profitability that would have been needed with the previous system and can therefore be seen as stores which the organization would not have otherwise developed. Let us further assume each new store generates a revenue of £5m per annum and has an average net profitability of 10 per cent. Table 10.2 illustrates the flows of profitability over this period of time. The total sum of £37.5m is over 100 times the total cost of the SDSS developed by us for the client. This gives a fairly clear indication of the substantial business benefits the client is receiving from investing in intelligent GIS.

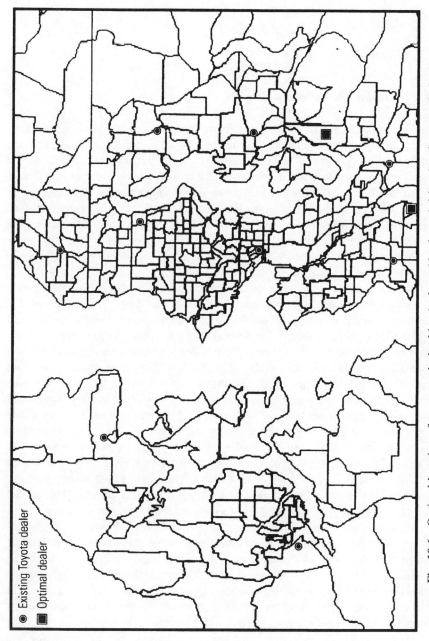

Fig. 10.6 Optimal location of two new dealerships (at least 10 minutes' drive time between dealers)

- ◉ Existing Toyota dealer
- ■ Optimal dealer

Table 10.2 The financial return on GIS investment over five years

Year	1	2	3	4	5
Benefits (£million)	2.5	5	7.5	10	12.5
Total = £37.5 million					

10.4.4 A major UK building society

The financial service organization that we have dealt with has a network of several hundred branches throughout the UK, although there are significant regional concentrations of the network. The particular issue that the organization was concerned with was the performance both of the brand and of the individual branches within the area bounded by the M25 motorway in South-East England, mainly in the London conurbation. The organization was considering investing several tens of millions of pounds in refurbishing the hundred or so branches in this region. It was believed that the poor performance of the branches and the brand could partly be explained by the relatively poor physical fabric of the branches.

However, before embarking on this programme, it was deemed appropriate to delve deeper into the real causes of the problem. This involved the construction of a national information system that contained detailed information on the demand for several different financial services products by small area across the country, estimated by marrying information on population types (age, sex, social class etc.) with their propensity to hold accounts of certain types (high income deposits, mortgages etc.). We also identified the location of every single bank and building society outlet and assigned these branches to some 2000 'financial services centres' over the whole of the UK. Using the organization's customer data it was possible to calculate the market share that it had achieved in each centre and the market penetration it had achieved in each small area for the whole product range. A spatial interaction model was designed in order to replicate the flows of account holdings between residential areas and financial services centres. This model was calibrated using the organization's own customer data in such a way that we accurately reproduced the behaviour of their own customers. Given that detailed competitor data were unavailable an assumption was made that the customer behaviour of competing banks and building societies for each product would be broadly the same. The full model therefore reproduced the consumer behaviour across the whole range of products for all competitors in the business sector. In this way we could estimate the relative performance of competing banks and building societies and were able to use this information to assess the actual performance of the organization that we were working with.

Once this detailed system had been implemented it was used to perform a 'critical success factor' analysis which attempted to tease out, from the information and the model output, which factors caused good performance in certain regions and which factors generated poor performance. It turned out that three main factors appeared to be at work. The organization tended to perform poorly when the following conditions occurred:

1. A low level of population per branch in a region.
2. A high level of representation of its main competitors in the region.
3. Relatively high regional house prices and average mortgage values.

Our analysis demonstrated that the significant investment that was to be made in refurbishing the branch network in South-East England would not address the main structural problems being experienced there. Furthermore we identified opportunities in other regions of the country where the critical success factors were in evidence but the organization had an underdeveloped branch network. Our recommendations were that the organization should not go ahead with its multi-million pound investment in refurbishment, but rather that a programme of network development be pursued in suitable regions, and that this might be coupled with the possible acquisition of a smaller financial service organization which had a suitable regional base to its operations. In this example the quantifiable benefits are not so much the profit streams to be generated by the development of the intelligent GIS, but more the significant savings that have been identified through not proceeding with their planned investment. The application also demonstrates that development of an intelligent GIS approach can provide key inputs at a senior level within an organization and not just assist with tactical operational issues.

10.5 Conclusions

The argument of this chapter has been that proprietary GIS is currently failing to address principal business objectives because of its lack of analytical power. We have argued instead for the incorporation of spatial models within GIS through the idea of spatial decision support systems. In particular, we have been keen to stress how SDSS can address key business objectives and the considerable financial benefits which can accrue. Given the nature of spatial model construction it is likely that such analytical power could not be provided in a totally generic fashion. Such generic solutions appear far more suitable to pattern analysis and clustering techniques where common problems exist. Business objectives are far more complex and demand solutions only attainable from sophisticated (and often uniquely calibrated) spatial modelling capabilities. We believe the solution lies firmly within customized SDSS. The development of customized SDSS will not necessarily replace simpler, cheaper packaged systems because they both address different market requirements. Nor do customized solutions have to be re-invented for each application: they can be built from a set of modular software and analysis tools. However, only through the development of decision support systems will the true potential of GIS to address strategic business issues be fully realized.

Part Four

Geography in Business?

A recurrent theme through much of the first three parts of this book has been the linkage of the business and service sectors to the world of academia through applied research. In Part Four we will consider these links more explicitly, with reference to experiences in the US and UK.

In Chapter 11, Nora Sherwood reflects upon the weaknesses of US geography in the school curriculum and the knock-on effects that this has for the subject at higher education levels. US society is technologically advanced, and the US continues to have a major role in the development of the global GIS industry. However, with regard to the development of applied research, these developments have to some extent at least passed geographers by. In Chapter 12, Clarke *et al.* begin with a discussion of the stronger education base for geography in the UK system, but nevertheless point out that geography education faces upheavals and potential problems in the near future. They then describe three recent UK initiatives which have sought to build bridges between academia and applied research in the area of business and service planning. They develop a typology of linkages and begin to draw lessons as to the kinds which are the most cost-effective, productive and enduring.

There are elements of continuity and change in all of this. With regard to the US, Sherwood in Chapter 11 detects some evidence of a renaissance in geographical thinking, which she views as ushering in a possible 'golden age' for US geography. She identifies the emergent area of 'business geographics' as a focus of interest in applied geography. However, most of the case studies that she cites conform to the established pattern of geographers participating in near-market research on only a part-time and piecemeal basis. Geography has long been viewed as less applied and less relevant than some other social science subjects (notably business studies), and it is difficult to see how this image might change based on these contributions alone.

There is also some continuity in the UK case studies in that the initiatives described in Chapter 12 can be viewed as similar in spirit to previous 'near-market' research projects, such as the early funding for the development of Super Profiles (as described by Batey and Brown in Chapter 5). However, the strongest currents are those of change, particularly with regard to the development of competitive and revenue-generating pressures in the UK university system since the 1970s. These have been quite different from anything that has gone before, and have stimulated the need to generate alternative sources of funding, such as externally funded research contracts. External contracts can and do entail work that is of intrinsic academic, as well as commercial, merit. However, the evidence that the contribution of such research is recognized by UK academic institutions is limited. Ultimately, perhaps, the contributions of geographers to business and service planning are likely to be determined by these institutional attitudes, as well as by the number and extent of niches within which geographers' skills may be exploited to mutual advantage.

11

'Business geographics' – a US perspective

Nora Sherwood

11.1 Introduction

The goals of this chapter are two-fold: (1) to examine the interactions between business and academia as they relate to applying GIS to real-world business challenges in the US; and (2) to review the status of the interaction between GIS and business in the US. The chapter is divided into two sections because in the US there exists relatively little interaction between business, geography academia and GIS. Therefore, to portray adequately the status of GIS in 'business and service planning' it is necessary to develop this element separately from the main theme of this book.

11.2 Business, academia and GIS in the US

11.2.1 Geography as the basis of GIS

In the rush to create bigger and better technical solutions, many in the GIS industry tend to forget that the discipline known as 'geography' is the basis of GIS. GIS provides nothing more than the opportunity to manipulate and analyse geographical phenomena using automated systems. In fact, Michael Goodchild, director of the US National Center for Geographic Information Analysis (NCGIA) quite recently suggested (Goodchild 1992) that the acronym GIS should be understood to stand for 'geographic information science'. This new definition would place more emphasis on analysis of 'geographic information' and less on 'system'. Therefore, to understand the status of business, academia and GIS in the US, it is necessary to first understand the status of geography as a discipline in the US. In general, there is good news and there is bad news. The bad news is that the status of the study of geography is poor. The good news is that it is improving rapidly.

In the UK, geographical study is popular. According to a recent editorial in *GIS Europe*, judging by entries for public examinations, geography is 'the most popular subject in UK secondary schools' (Roper 1993). While many in the UK urge improvement in the quality and quantity of geographical education, there is no question that it is more advanced in most respects than in the US (Department of Education and Science 1991; Rawling 1992; but see Clarke *et al.*, Chapter 12, this volume). Because of my diplomatic upbringing, I have had the opportunity to compare English- and American-dominated educational systems, and my personal experience confirms what many believe to be the case: the US lags behind the UK in both quality and quantity of geographical education at all levels.

This fact carries through to the interaction between business and academia. However, before understanding why this is the case, it is first necessary to delve into the status of geographical education in the United States.

11.2.2 Status of geography and geographical education in the US

Geography as an academic discipline has had high points and low points during its brief history in the US. Geography has been taught in this country since before independence; elements of geography formed parts of the curriculum of Harvard University since the 1600s (Stoltman 1990). For the most part, the early curriculum focused on rote memorization and description of physical geography. During the early 1900s the focus shifted to cultural geography. Then later in the century geography began to lose its individual identity along with other disciplines as they became categorized as 'social studies'. This loss of identity began a period of decline in geographical knowledge, which has been well documented (Gandy 1960; Adams 1960; Mayo 1964). While geography and sister disciplines such as planning continued to be small, but important and respected, academic disciplines within universities in the US, the period between the mid-1940s and the mid-1980s could be considered the 'dark ages' of geographical education in elementary, junior and high schools. The status of geographical knowledge in the US has become cause for national embarrassment. The popular news media take great pleasure in publishing 'facts' such as '92.6 per cent of US school children can't find the US on a map of North America'. While perhaps indulging in a bit of exaggeration, the point is that Americans are not known for their strong knowledge of global geography (National Geographic Society 1988).

There are probably many reasons for the lack of emphasis on geographical knowledge in the United States. Because of the economic and political environment of the past decades, the US has had a tendency to 'look inward'. It has been a military, economic and political 'superpower', and therefore there has not been much reason for the average American to think about the rest of the world. To better illustrate this point, consider the following economic example. California's economy is ranked among the top ten largest economies *in the world.* An economy that large, within a country, with relatively low barriers to entry for US companies, provides a disincentive for US companies to look beyond their own borders for trading partners. This is also evidenced by the lack of teaching of foreign languages. Of course, this situation is now changing rapidly, and the resurgence of the importance of geographical education at all levels is gaining strength.

During the early 1980s, while I was an undergraduate student in the Department of Geography at the University of Colorado, the shift in the importance of geographical education began. The State of Colorado became one of the first to institute a geography requirement. If students did not study geography in high school, they were required to enrol in geography classes in college in order to graduate. When I graduated in 1984, I was in a class that numbered in the tens. Current geography classes at the University of Colorado number in the hundreds. This is clearly attributable to the mandatory exposure to geography as a discipline in either high school or college in Colorado. According to the 1993–94 National Survey of Course Offerings and Testing in Social Studies report (National Council for Social Studies 1994), 28 states and the District of Columbia now have geography requirements in the K-12 programmes.

In the early 1990s, President Bush and the nation's governors jointly adopted the National Education Goals; five core subjects were identified as crucial (US Department of Education National Assessment Governing Board 1993). The core subjects were English, mathematics, science, history and geography. Goal 3 of the National Educational Goals states: 'By the year 2000, American students will leave grades four, eight, and twelve having demonstrated competency in challenging subject matter including English, mathematics, science, history, and geography; and every school in America will ensure that all students learn to use their minds well, so they may be prepared for responsible citizenship, further learning, and productive employment in our modern economy.' This was a clear sign of geography re-establishing itself as a separate entity within the national educational environment in the US.

There is much other evidence of this resurgence, including the establishment of national educational goals and standards for geographical education in primary and secondary schools, and the National Geographic Society-sponsored Geographic Alliance Network (National Council for Geographic Education 1993; Business Geographics 1993). The Alliance began educating teachers in 1986, when fifty teachers attended the Alliance Summer Geography Institute (SGI) programme. In 1993, 1572 teachers attended a variety of Alliance-sponsored institutes, including the SGI, the Instructional Leadership Institute, the Educational Technology Leadership Institute and the Workshop on Water. The focus of these institutes is not only to teach teachers how to teach geography, but also to teach teachers to teach other teachers to teach geography. The National Geographic Society estimates that, through in-service workshops, the programme has probably reached about 45 000 teachers, who in turn have reached almost 3 million students (National Geographic Society 1994).

The GIS community itself is helping to develop geographical awareness in schools. For example, the Environmental Systems Research Institute (ESRI), the company which sells ARC/INFO and other GIS software products, sponsors a programme called 'Adopt a School'. The programme provides free software and data to schools to help raise levels of geographical awareness.

11.2.3 Impact of status of geography as a discipline on the business community

While it is clear that geographical education is currently experiencing an upsurge in the US, this has had little effect on the current generation of professional geographers and their interaction with 'the rest of the world'. The 'baby boomer' generation, born between 1946 and 1964, and the generation just before them, were educated during the 'dark ages' of geographical education in the US – approximately 1940 to 1980. These are the people who are primarily in control of the US's 21 million businesses. As previously suggested, geography as an identifiable discipline lost its identity during this time. This is evident in that outside academia, almost no one would consider themselves a 'professional geographer'. In fact, in 1988, 44 per cent, or close to half, of the membership of the Association of American Geographers (AAG) held positions in college or university teaching (Stoltman 1990). At that time, there were 5903 members. That means that there were only approximately 3000 others who associated themselves with the profession to the extent that they enrolled as members of the premier professional society. When student membership is accounted for, there were probably very few remaining.

This should not be taken to imply that there are not individuals who are professional geographers outside academia in the US. What it does mean is that they are not visible. According to Stoltman:

> Geographers are often engaged in teaching, researching, or practising urban and regional planning, but not always under the title of geographer. Similarly, environmental studies programmes and projects usually have an interdisciplinary focus, but geography and geographers invariably provide significant scientific information to such programs. Neither is geography, nor are geographers, often highlighted as major contributors to those endeavours in the eyes of the public. Similarly, public lands and natural resource use and conservation issues are within the domain of geography, but credit is not often extended to the discipline for its contributions.

This is a symptom of the lack of identification of geography either as a profession or as a discipline. My experience at the University of Colorado provides personal experience of this phenomenon. The Department of Geography actually confers two degrees, one in geography and another in 'environmental conservation'. The core curricula for both are very similar.

This rather long-winded explanation of the status of geography and geographical education will, I hope, help readers outside the US to understand that when the business community in the US seeks help in applying GIS to its business challenges, the local university's geography department is unlikely to spring to mind as a source of help! Abler blames this status on geographers, and alienation from the practical world on themselves: 'geographers have allowed a gulf to open between theoretical and applied science, to the detriment of both' (Abler 1993). While perhaps a simplification of the current situation, the comment is a good summation of the status of the integration of academia and business in this country.

The complex issues described previously regarding the status of geography and its integration with business in the US are complicated yet further as they relate to GIS. The academic environment in the US is currently involved in a 'love–hate' relationship with GIS. According to Abler, 'no more than half the academics and practising geographers in North America exhibit any interest in GIS'. In fact, some exhibit outright hostility. In many academic departments, there is quite a struggle going on between the 'GISers' and the 'non-GISers'. A few of the more radical faculty members would prefer that GIS not be taught in geography departments. This attitude is reminiscent of the consternation caused within geography departments by the 'quantitative revolution', when it was discovered during the 1950s and 1960s that statistics and numerical analysis could be applied to geography. Like that earlier turmoil, except for continued resistance by some greybeards, GIS will eventually be accepted into the hallowed halls of US academia. However, at the moment the struggle is still very much alive.

Retail modelling provides a very good example of this struggle in the academic community, as well as the more general struggle to integrate geography and business. Retail modelling has been one of the key contributions of US geographical academia to the business community (Applebaum 1965; Huff 1963). At one level, these models are rather complex, and require significant resources to implement in a manual mode, meaning that 'relatively few organizations have benefited from their existence' (Daniel 1993). At another level, academia has been slow to integrate these models and GIS. For example, the 1993 annual AAG meeting included no papers on this topic. In so far as this integration is happening, it is primarily being spearheaded by the business sector, a good example of which is Tydac's early integration of gravity modelling into the SPANS product line.

11.2.4 What about the planners?

Planning is a discipline that is closely related to geography. Land expanses are vast in the United States, and land planning has been stressed much less than in places like Europe. Nevertheless, local jurisdiction planning is an important academic and applied discipline in the US. Local jurisdiction planning applications continue to be some of the most logical applications of GIS, and many of them were originally conceived by academics in the planning discipline. Perhaps the quintessential 'GIS planning application' is using a buffering capability to produce a list of properties within a specified distance from a property requesting a zoning change. However, planning in the US is relatively under-funded and has difficulty getting the money needed to purchase software, data and hardware to perform such applications. The Urban and Regional Information Systems Association (URISA) is perhaps one of the premier GIS associations in the US, especially as it relates to government. During the last several years, URISA has evolved from seeing its target member as a local jurisdiction planner who is using or could use GIS, to seeing that member as a local government information systems professional.

Because of the lack of funding, many software vendors in the US see city or county planning departments as secondary or tertiary markets for their products within a local jurisdiction: engineering and public safety, for example, are much better targets because of better funding. While their need for GIS technology may not be as 'classical' as that of the planners, they do have tremendous spatial challenges that can be addressed using GIS. Engineering departments often receive project-specific funds that allow them to purchase GIS technology. In many jurisdictions, the engineering department may include a revenue-generating entity such as a public utility, giving them even more opportunity to purchase GIS technology. Public safety is often a well-funded entity within local jurisdictions, because perceived and actual increases in crime within communities leave residents more inclined to support increased taxes for public-safety purposes.

Several summary conclusions can be drawn regarding academia, business and GIS:

1. Geography has not had a significant or defined role in the academic upbringing of our current generation of US 'captains of industry'.
2. There is no strong connection between academic geographers and business.
3. Some academic geographers are somewhat ambivalent regarding GIS.
4. Academic geographers have played only a limited role in the private sector's rapid movement towards GIS.

A goal of this volume is to identify where links are being and have been forged between business and academic geography as they relate to GIS. Having described why there is so little integration between business and geography in general, and between business, geography and GIS specifically, it would be easy to simply give up writing this chapter. However, while these are general trends, there are many individual projects in the US where there is indeed a very solid integration among these three components. Some of these will be described in the next section.

11.3 Integration between business, geography and GIS

In this part of the chapter, we will look at three categories of integration between business, geography and GIS:

1. Business projects in which a geography department is involved.
2. Business projects in which former geography students are involved.
3. Business projects in which a consultancy of professors is involved.

To fully understand integration between geography and business, it is important to take a look at each of these categories individually.

11.3.1 Business projects and geography departments

Perhaps the best example of an excellent project in which a geography department and business entity are working together is the South Carolina State Development Board and the University of South Carolina. The project was awarded the Industrial Development Research Council's Outstanding Area Research Award and has been shown on the popular US television programme 'Good Morning America'. As a graduate student at the University of South Carolina in 1986, I had the opportunity to be involved with the very early stages of this project.

The economic development programme is run by the state, and is charged with 'marketing' the state of South Carolina to businesses in order to convince them to locate in the state. At a first glance, this project might appear to be one in which a university and government agency are working together, and while that is the case, the emphasis is clearly on the business aspects of attracting industry. In this country, states vie with each other and there is strong competition to bring an automobile manufacturing plant, for example, into a state. This is particularly true in the south-east, a region that is still recovering from the exodus of the textile industry to foreign shores, and, some would say, from the Civil War! States spend millions of dollars to promote their work force, economic conditions and quality of life. Dr Dave Cowen, a professor of geography in the University of South Carolina (USC) and Sena H. Black, who was (formerly) with the Development Board, wrote the following illuminating comments about university GIS programmes and their general ability to provide service to the 'real world':

> Throughout the history of GIS, universities have played an important role in the development of technical aspects of the field. They also have performed innovative research and have served as the training ground for much of the manpower for the field. At the same time, they are notorious for their poor performance as sites for supporting operational GIS activities. Nevertheless, all the participants in the design and implementation of the [South Carolina Infrastructure and Economic Development Planning Project] system acknowledge that the key to its success has been the relationship between the Development Board and the University. (Black and Cowen 1991)

Despite this condemnation of university support of operational GIS, the University of South Carolina has done very well. The SCIP project (South Carolina Infrastructure and Economic Development Planning Project), as it is known, is a major state-wide GIS. It includes databases that can be used primarily for site selection. Data layers include: transportation; water supply systems; waste water systems; air quality; land cover and land use; flood plains; demographics; directories of existing businesses and industries; available buildings; and available sites. These databases support queries such as 'Find an available building that has at least 300 000 square feet of space suitable for manufacturing, has railway access, has sufficient water supply to support activity *X*, is near to the suppliers of materials needed for activity *X*, and is surrounded by a workforce with *X* characteristics.'

In addition to responding to this kind of query, the system offers the much vaunted advantage of graphics and interfaces which give the impression that South Carolina must be a very progressive, forward-thinking state to have such a system – in other words, that it must be a good place to locate, relocate or develop! While the state has been in the position to reap the most visible benefits, USC also has gained significantly from the partnership. The richness of the database developed for the project has allowed USC to be involved in other projects with local utilities and with a NASA grant. In addition, the attention they have gained from the project has resulted in significant hardware donations (Black and Cowen 1991: see Clarke *et al.*, Chapter 12, this volume, for an account of similar donations made in the UK context).

11.3.2 Business projects in which former geography students are involved

This section describes two projects in which students from USC's GIS programme are involved. While a USC bias could be argued, the fact remains that it has been an important institution in providing a solid GIS background with a very applied focus. The first project is the Georgia State economic development programme and the second is a major GIS implementation of the United Parcel Service (UPS).

The GIS project manager at the South Carolina State Development Board was William Shinar, a former graduate student of Cowen's at USC. He recently moved over to the University of Georgia to work on that state's economic development programme as the state's Economic Development Director. Unlike USC, which is primarily a teaching institution, the University of Georgia is the state's 'land grant' college. A land grant college's traditional goal has been to transfer technology from its research programmes to the state's population, with the objective that this will improve the state population's quality of life. (The most typical and traditional example of technology transfer is in agricultural programmes.) As such, the University of Georgia has three primary divisions: teaching, research and service. The service division is where the technology transfer takes place. Shinar is in the service division and is not attached to a geography programme at the university. While South Carolina's economic development programme is somewhat centralized within the development board, Georgia's programme is more decentralized. The work Shinar is doing is primarily focused on developing databases to support decentralized economic development groups. These include the state's

Department of Industry, Tourism and Trade, Georgia Power, Oglethorpe Power, Savannah Power and Light, the Georgia Resource Center, various local and regional chambers of commerce, and small business development offices. At the time of writing (1994), the programme is still in its first year. Shinar and his team are close to finishing cleaning the street file databases, which will serve as the transportation base layer. In addition, they are mapping existing industries across the state, as well as available buildings and sites, water and sewer systems and traffic flows.

United Parcel Service (UPS) is a major package-delivery service, delivering 11 million packages daily in 185 countries. Tony Lupien, GIS Division Manager at Roadnet Technologies and former USC student, has been instrumental in supplying GIS-related technology to this giant organization. Roadnet is a wholly owned subsidiary of UPS. During its 85-year history, UPS has 'depended on industrial engineering to optimize delivery, distribution, logistics activities and movements' (Kunze 1993). The company has managed to cut costs and time in delivering packages by analysing all the elements that play a role in package delivery. In 1985, they began to analyse GIS as a possible tool to increase efficiency. Because of competitive advantage issues, the company is somewhat cagey about describing exactly how GIS is exactly used. We do know, however, that GIS is used 'for day-to-day operations including package sorting, routing, truck loading and dispatch' (Kunze 1993). GIS is also used in the company's On-Call Air service, in which a customer may call an 800 (toll-free) number to request package collection. In addition, UPS has formed a new company, UPS Worldwide Logistics, to make their GIS knowledge available to UPS customers.

It is clear that students coming out of GIS programmes are making substantial contributions to the implementation of GIS in business and service planning. It has only been in the last ten years that such students have been produced by programmes around the country, but it is clearly a strong trend. According to *GIS World*'s database, there were, in 1994, 580 college and university programmes in the US and Canada that taught at least one course in GIS. Randy Majors is a good example of the type of student currently being produced by the University of Colorado. He is working for a commercial and industrial real estate brokerage whose owner has developed a substantial manual and automated GIS to assist in 'making real estate deals'. Majors' objective is to develop a data sales programme to sell the brokerage's (non-proprietary) databases to other GIS users.

11.3.3 Business projects in which a consultancy of professors is involved

For the most part, universities in the US tend not to be entrepreneurial organizations, and do not adapt quickly to change (cf. Clarke and Clarke, Chapter 10, this volume; and also Clarke *et al.*, Chapter 12, this volume). This is especially true of arts and science colleges. In some institutions of higher learning, real-world applications can even carry a stigma because they are not considered 'academic', which by definition is 'not applied' (cf. Openshaw, Chapter 7, this volume). Because of this and other factors, quite a few professors have their own consultancies that apply GIS to business problems. In some cases, these consultancies use techniques that were developed as part of a professor's research. This section describes three such consultancies: Bowne Distinct, Matrix Research, and GRAASroot. Like many consultancies that work with private-sector clients, all three are somewhat evasive about naming and/or describing their clients, but are happy to describe techniques used.

Bowne Distinct is a logistics company based in Columbia, Maryland. Dr Larry Bodin, who is a professor of operations research at the University of Maryland business school's Management Sciences and Statistics division, is one of its principals. Research at the university is orientated towards developing test algorithms that solve simple problems on small databases. These algorithms might, for example, work towards solving routeing problems with barriers such as one-way streets. The algorithms developed at the university are then transferred to Bowne Distinct for fuller implementation in real-world cases. The routeing algorithm that addresses specific issues such as one-way streets would be further developed and revised to address an actual client's needs.

Dr Brady Foust of the University of Wisconsin, Eau Claire, and Dr Howard Botts of the University of Wisconsin, Whitewater, are both professors of geography. They are partners in a GIS consultancy called Matrix Research. Their primary interest is in developing and analysing actual and potential trade areas for retail clients (Foust and Botts 1993). They have developed a technique to create 'market penetration polygons' (MPPs) that come closer to exactly portraying a store's trade area demographics than do traditional 'ring' studies (cf. Birkin, Chapter 6, this volume; Cresswell, Chapter 9, this volume). Foust and Botts use a variety of techniques to establish an MPP, including traditional licence plate surveys. Licence plate surveys were pioneered in the US by R. L. Polk during the 1970s. This technique consists of going to a store's or a shopping centre's parking lot and writing down a list of licence plate numbers of all the cars found in that parking lot. Each car owner's address is then extracted from the state's Department of Motor Vehicles registration files. In almost all states in the US, this information is in the public domain and available at a relatively low cost. Once the address is obtained for each car, those addresses can be plotted on a map (now done using computers, street files and address geo-coding techniques) and the rough trade area can be established for that store or shopping centre.

Dr Michael Robbins is the academic behind GRAASroot Real Estate Counseling, a Denver-based consultancy that works with real-estate issues. Robbins is a professor at the University of Denver College of Business Administration's Real Estate and Construction Management Program. His consulting practice has been involved with projects in Colorado and elsewhere that dealt with valuation, site selection and development. Robbins said in a recent interview: 'I like to believe that most of what I do is walking the tightrope between business and academia' (Robbins 1993). He uses scholarly research to find real-estate applications that can only be addressed using GIS. Then he works with these applications for real-world clients in his consulting practice.

Although there may be relatively less direct interaction between academia and business in the US than there is in the UK as it specifically relates to GIS, the previous case studies provide examples of how indirect interaction has and will continue to affect the 'business geographics' market evolution.

11.3.4 Future opportunities

As the status of geographical education in the US improves, there will be numerous opportunities for the geographical academic community and the business GIS-user community to forge ties to work together. In the last year I have spoken to many in academic organizations who are interested in developing components of their GIS programmes that will specifically focus on 'business geographics' applications. In addition, business academics are rapidly discovering GIS and implementing it within their own programmes. *GIS World* has published a book, *Profiting from a Geographic Information System* (Castle 1993a), which has been adopted for several classes. It was certainly not designed to be used as a class text. This somewhat surprising development clearly indicates that some GIS programmes are starting to include practical business applications in their curricula.

11.4 Status of 'GIS and business' in the US

Regardless of the relative lack of participation by the geographical academic community, the business community has taken hold of GIS technology. We are attempting to adopt a separate name for these activities: business geographics. The rationales behind developing a new name are several-fold:

1. It uses the word 'business' and thereby helps to identify a set of tools with a user group.
2. The business industry did not seem to be catching on to the acronym 'GIS' and so clearly something else was needed.
3. There are significant enough differences in how applications are structured to warrant a separate name.
4. It is slightly ambiguous, and is therefore a wider 'umbrella' for a variety of types of geographically related technology that might not be included in a strict definition of GIS (e.g. global positioning systems, scanned maps and multi-media interfaces, consumer travel 'toys', etc.).

The way this name came about is interesting. A *Forbes* article, written about GIS technology, used the title 'Geographics'. This sparked debate in the industry, particularly in the pages of *GIS World* magazine, about whether someone from outside 'our' industry had the right to rename it. In the end, as it became noticeable that 'GIS' simply did not connect with potential new users, it became apparent that a new name was probably appropriate.

'Business geographics' applications are significantly different from what we call 'traditional GIS' applications (Sherwood-Bryan 1993a). These traditional applications tend to be used by government and utilities, as a broad generalization. Business geographics applications have three key character-istics (Sherwood-Bryan 1993a):

1. Application focus. The 'generic toolbox', with some 2000 functions, is replaced by tailored applications in which the application vocabulary, flow and functions, and data accessed, are tightly controlled. The applications tend to be more focused on doing one or several specific things, rather than providing all capabilities for every potential purpose. This increases the level of user-friendliness, still a major concern for many users of GIS.
2. Data development. In general, business geographics users are not data creators, whereas many traditional GIS users are. Governments and utilities are often involved in the process of digitizing data as part of developing and using a system. It is unlikely that a business geographics user would digitize maps; it is much more likely that the data used for the targeted application would be bought from an off-the-shelf source.
3. Data focus. While traditional GIS users have often been land managers (roads, utilities assets, etc.), business geographics users generally are not. The geographical data they have in their systems are used as a back-drop for the data they are really interested in (customers, sales, demographics).

This name is beginning to catch on. We have noticed several consulting companies beginning to use the phrase to describe the services they offer, and it is starting to show up in sources other than *Business Geographics* magazine. Most noticeably, a new company in the UK has quite recently (September 1993) announced that they are using it for their own company's name.

11.4.1 History and trends

Business geographics is currently *the* fast-growing worldwide market segment for GIS software. Recent research indicates that between 1990 and 1991, the growth in software used for business geographics was 126 per cent (GIS Strategies 1993a). While still a relatively small component of the software market (8.6 per cent in 1993 for a total of $60.2 million), it is expected to continue to grow quite quickly and to be the third largest consumer of software by 1997 (14.6 per cent for a total of $192.3 million), behind utilities and government (GIS Strategies 1993b). When considering that there are 21 000 000 businesses in the US alone, these estimates could be conservative.

The growth of attendance for the *GIS in Business* conference series has been equally rapid. The first conference, held in Denver in 1992, attracted 1100 participants. The second conference, in Boston in 1993, attracted 2200 participants. The first European conference took place in Amsterdam, in January 1994, when approximately 1100 people attended, which is close to the same number who attended the first US *GIS in Business* conference. The second European conference is taking place in Madrid in 1995. The raw attendance figures lend validity to the notion that business users are different from traditional users. In addition, as an attender, I found that the European business-data market is evolving more rapidly than I had expected.

Many in the industry believe that GIS is new to the business sector. This is not the case. While growth is extremely strong just now, there has been steady growth in the 'early adopter' group for some years. Early adopters, as described by Moore (1991) are those who are the first to try out a new technology. As early as the late 1960s, companies like Donnelley Marketing and R. L. Polk were developing and using rudimentary address coding guides. In the early 1970s, basic geographical 'ring studies' were used by early adopters. In the mid 1970s, R. L. Polk developed the methodology for associating licence plates to motor vehicle registration files so that trade areas could be developed and analysed. They sold (and continue to sell) these services to many large organizations throughout the country. The industry is still primarily focused on early adopters.

The future will bring continued fast growth. Other trends include:

1. Embedding. Already applications are being developed that cannot easily be defined as 'GIS' in the traditional sense. The geographical analysis tools are so deeply embedded within a system that users will not be aware they are using them. An example is the new Lotus One Source product from the Lotus Development Corporation, the same company whose Lotus 1-2-3 spreadsheets revolutionized numeric data handling. Lotus One Source is a PC-based business data product on CD-ROM that provides information on public and private companies, banking institutions and insurers. The accompanying application allows users to 'grab' data and map it by hitting a map icon, much as users can hit the graphing icon to create graphs.
2. Multimedia. There is increasing combination of the use of GIS and multi-media techniques. Several oil companies are scanning maps and attaching information icons that allow data and imagery retrieval much as does a GIS (Stover 1993). The differences are that the maps are not registered to real-world coordinates, and perhaps what is more important, they are 'dumb' images, lacking topology, layers, objects, etc.
3. Value added resellers (VARs). In the US as elsewhere, VARs are playing an increasingly important part in providing geographical technology to the business community, particularly given low levels of cost recovery on public-domain data-sets such as the US census (cf. the European case described by Waters, Chapter 3, this volume). These VARs generally tend to be more knowledgeable about a specific industry, such as insurance or health care, than is a GIS vendor. VARs are developing application-specific menus and screens and providing customized training and support. MapInfo, for example, has approximately 350 VARs.

4. Travel toys and kiosks. Some geographical technology vendors are taking advantage of travellers' growing interest in using computerized travel guides and atlases on notebook computers or personal digital assistants, and on information kiosks in hotels, airports and similar locations. Numerous products are becoming available, including inexpensive routers, trip planners, tourist guides and CD-ROM-based atlases.

11.4.2 Users

GIS World, Inc. recently received 442 surveys from readers of *GIS World* and *Business Geographics* magazines. Because of the way the results were structured, we were able to derive rudimentary comparisons between all responding users and business geographics users (Sherwood-Bryan 1993b; 1994). The following are some rough comparisons between the two groups of survey respondents, as shown in Figure 11.1.

The *Business Geographics* users who responded were more likely to use DOS, OS/2, Windows or Macintosh operating systems, but only half as likely to use UNIX (22 per cent versus 11 per cent). They were also more likely to consider 'ignorance about GIS and its uses' a major impediment to the growth of the industry (40 per cent versus 30 per cent). The *Business Geographics* survey respondents were slightly more likely to buy a product because it was inexpensive (14 per cent versus 12 per cent: GIS Strategies 1993c). Their pricing sensitivities were reported to be more intense. For example, 49 per cent of the group believed that users are only willing to pay up to $10 000 for hardware and software for GIS, versus 35 per cent for all responding users. Forty per cent said they would expect to pay less than $1000 for digital data, compared with 30 per cent for all responding users. The *Business Geographics* respondents were more likely to have purchased an existing geographical database (37 per cent) than to have created one (28 per cent). They were also less likely to have been involved in an in-house data conversion project (26 per cent versus 37 per cent). These figures suggest that the *Business Geographics* user group, at least as portrayed by the 111 respondents, varies not only in the reasons they implement geographical technology, but also in how they implement it and what they think about it.

11.5 Case studies

This section will briefly review case studies that highlight how business geographics is being used by various industries in the US. The selected industries, banking, insurance, retail, restaurants, consumer products, telecommunications, publishing and real estate, are by no means the only industries using geographical technology but have been chosen because of the wide range of applications issues that they raise. In addition, only one or two

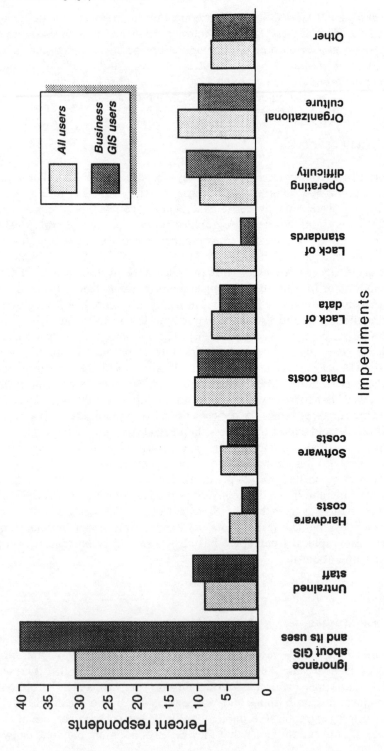

Fig. 11.1 Impediments to GIS adoption cited by business and other GIS users

separate applications are described for each industry, which is far from exhaustive. Because much of this book describes case studies, in this section I attempt to address applications that would be unique or somewhat different in the US than elsewhere.

11.5.1 Banking

Perhaps the most important use of GIS in the banking industry is regulatory compliance (Tavakoli A. 1993). The Community Reinvestment Act (CRA) and the Home Mortgage Disclosure Act (HMDA) are both laws which were passed in order to ensure that the banking industry does not practise 'red-lining'. Red-lining is the illegal practice of geographically discriminating against groups (primarily minorities) by refusing to meet the credit needs of customers. For example, if a bank has a branch in an area and accepts deposits from customers within that area but refuses to lend money to those customers, that is considered red-lining. Red-lining reporting is done at the census tract level, making it an ideal application for business geographics. Banks use GIS to analyse whether they are guilty of red-lining, and if not, to help prove to regulators that they are not. Theoretically, if they do find that they are guilty of red-lining, they can use GIS to market to specific groups of customers in order to increase their lending activities within an area: I say 'theoretically' because I have not heard of a bank that has admitted to using GIS this way since to do so would also be to red-line in a different way, which would also be against the law.

Another key application for banks has been to analyse mergers and acquisitions (Tavakoli H. 1993). To state a complex situation very simply, capital requirements for banking institutions have grown in the wake of the US savings and loan scandal, in which many savings and loan institutions failed because they were undercapitalized, or in other words, could not cover their debts. This has caused unprecedented consolidation in the industry as banks band together in an attempt to meet new capital requirements. In looking for merger or acquisition partners, it is important to be sure that combining with the target bank will improve the bank's situation, not make it worse. GIS is playing a role for banks who are looking for merger partners. Key questions include: (1) how well is the target bank complying with statutory requirements; (2) how does the target bank's retail distribution channel (banks and automated teller machines) complement the existing channel; and (3) how healthy is the target bank's customer base, based on the demographic characteristics of its catchment neighbourhoods.

11.5.2 Insurance

It is highly likely that the insurance industry will soon face similar regulations to those already faced by the banking industry (Mertz 1993a). The reporting is

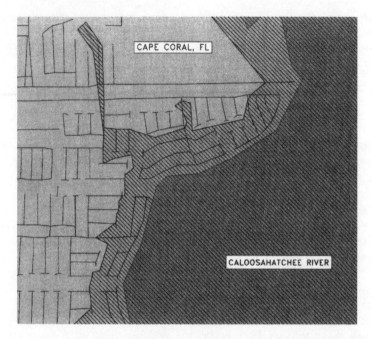

Fig. 11.2   GIS for insurance risk assessment, Caloosahatchee River, US
(grey: low risk; hatched: high risk)

likely to be at the zipcode (postcode) level. However, since this is in the future, this section on insurance will focus on current applications: risk assessment and avoidance. In the US, natural disaster after natural disaster has occurred over the last four years, bringing the total insurance industry's bill to $34 billion (Mertz 1993b). Because of this, the industry has begun seriously to consider how it underwrites policies. The Oakland fire, Hurricane Andrew and the floods and other catastrophes of the summer of 1993 each pointed out the need for greater underwriting care. Insurance companies are beginning to realize that they must either refuse to insure properties that are at great risk from natural disasters, or justify larger premiums for doing so, or clearly identify precise areas in which property damage may be partially reimbursed from other sources. For example, the cross-hatched areas in Figure 11.2 are wind-pools: insurers who write property-damage policies in these areas may be partially reimbursed by a state fund.

Analysing properties in relation to the geographical likelihood that they will be damaged because of a natural disaster is not possible without using geographical analysis (Runnels 1993). These kinds of application are good examples of 'embedding' because they primarily depend on a property being compared with a 'geo-file', or a listing of addresses that has been coded for likelihood of damage using GIS. In this instance, operators are unlikely to see a map during the underwriting process.

11.5.3 Retail

Levi Strauss & Company (LS&Co) is a good example of a product retailer that has begun to use geographical technology to customize 'product mix', or the combination of products available in specific stores (Allen 1993; see also Cresswell, Chapter 9, this volume). LS&Co is one of the world's largest clothing manufacturers, and sells many product lines in addition to Levi's jeans. Unfortunately, the company's sales had been lagging significantly, primarily as a result of mergers, acquisitions, price wars and significant retailers such as *The Gap* chain creating their own product lines instead of selling LS&Co's products.

'Micro-marketing' has become a popular term for the strategy which LS&Co used to turn their poor sales performance around (again, see Cresswell, Chapter 9, this volume). Micro-marketing consists of paying much closer attention to product marketing and product mix, at a very fine geographical level. A highly visible example of micro-marketing would consist of using Spanish-language billboards in a Los Angeles neighbourhood that is primarily Hispanic. In LS&Co's case, micro-marketing consisted of tinkering with the product mix on a store-by-store basis. Since LS&Co primarily sell their products through distributors other than themselves (i.e. they own very few stores themselves), this means they must achieve a high level of coordination with their retailers, from large department stores to small boutiques. Their goal is to help the retailers sell large quantities of their products. By geographically analysing the demographics within a specific large department store's trade area, they can begin to make assumptions about the kinds of product that would be appropriate for that store (Santoro 1993). For example, a slightly older and more affluent population would be likely to warrant more of the 'Dockers' products such as khaki slacks, versus a younger crowd, who would buy core Levi's products such as jeans and jeans jackets.

11.5.4 Restaurants

Among fast-food chain restaurants such as McDonald's and Burger King, the most visible and vocal users of GIS are at Arby's. Among a variety of applications used at Arby's is one that uses drive time to establish the likely trade area for an existing or potential store (Freeling 1993). In the fast-food business, customers are likely to be attracted more to the convenience of the product than to its gourmet appeal. It is, therefore, important to look at a trade area from the perspective of drive-time accessibility. Like many retailers, Arby's personnel know how long someone will drive to access their product. In addition, they are familiar with the demographic characteristics of their typical customers. By analysing the demographics of a trade area established by using drive times, the likely sales performance for a store can be modelled. Arby's is careful not to 'cannibalize' an existing store when developing a new store. Cannibalizing means taking customers from one of their existing stores (see also Cresswell, Chapter 9, this volume, for a UK perspective). Cannibalization should be avoided since a new store should increase overall business, not spread it around. By the same token, they are very eager to take away their competitors' customers, and so they carefully analyse their competitors' existing locations in comparison with their own.

11.5.5 Consumer packaged goods

Consumer packaged goods are small, non-durable items that are generally bought at the grocery store. These might include soda, breakfast cereal and laundry detergent. Like most retailers and manufacturers of consumer goods of any type, the consumer packaged goods industry is beginning to use micro-marketing techniques (Buxton 1993). Through using a sophisticated combination of databases and manipulation techniques, product marketing programmes can be developed at the retail chain level. In the US, almost all grocery stores use check-out scanners. These scanners read a bar-code on each product and automatically provide the price of the product. This is convenient for stores because they can change the price of a product (such as for a sale) without having to re-tag each item. Not only is this convenient for stores, but because of this process, a very rich database of what brand's products are purchased at what stores is generated. These data are aggregated into approximately 50 'scanner markets' that cover various portions of the US. By combining these data with the demographics that tend to drive product demand, such as age and income, a buying power index (BPI) can be modelled for stores. The BPI indicates how much of various products could be sold at that store. By knowing a store's BPI, the retailer can improve the product mix (micro-marketing) to 'push through more product' – to use the industry jargon.

11.5.6 Telecommunications

Telecommunications is currently one of the most dynamic industries in the United States and worldwide. Competition for the cellular telephone services market is strong in the US, and the company that can provide the best service has a significant advantage (Roan 1993). Current technology primarily relies on equipment within a 'cell site' to control the communications interface between other cell sites and the traditional telephone system. Because the location of each very expensive cell site determines coverage, which in turn determines the level of service offered, optimal location of the cells is critical (Sherwood-Bryan 1993c).

The industry is turning to geographical technology to find optimal cell sites. Geographical considerations include ensuring even and adequate spacing, providing service along well-travelled corridors, and analysing the cell site's 'view' based on a location's view shed. In addition, once a new geographical area has cellular service available, the telecommunications industry can use geographical technology to target-market the new product to the cellular telephone users within that area.

11.5.7 Publishing

The publishing industry has been quick to adopt geographical technology, particularly for selling advertising. The *Chicago Tribune* used GIS to realign its sales territories recently, a case study that provides a good example of how publishers are using GIS to improve and increase advertisement sales (Burkhart 1993). Free-standing inserts (FSIs) are a typical form of newspaper advertising. They consist of loose pieces of paper or booklets that are included in the folds of a newspaper. They typically advertise consumer packaged goods. Because of the cost of inserting FSIs into newspapers, good micro-marketers want to place them in only those newspapers that will be read by likely buyers of their product. Newspapers can provide an attractive value-added service by helping an advertiser select sales territories in which to include FSIs. The *Tribune* had been using the same, *ad hoc* sales territories for many years. The project in which GIS was used consisted of realigning the sales territories to conform with zipcode (postcode) boundaries. Using zipcode boundaries was strategically clever since the type of demographics that can be used to model demand for a product can be readily purchased at the zipcode level. (See Martin and Longley, Chapter 2, this volume, for an exposition of the principles guiding such marketing.) Because of the realignment, it became easier for the *Tribune* to communicate the potential that FSIs would have when placed in their newspaper.

11.5.8 Real estate

While the real-estate industry is most closely associated with location, it has been one of the slowest to catch on to the potential (Sherwood-Bryan 1993d). Nevertheless, it has started to implement a variety of applications (Castle 1993b). One of the most obvious, and least implemented, applications is supporting Multiple Listing Services (MLS) (Castle 1993c). An MLS is the tool that almost every residential realtor (estate agent) uses to analyse available properties. It is a computerized system that lists available properties and includes the characteristics of each property including size, type, number of bedrooms, listing price, etc. We are beginning to see inclusion of mapping capabilities into these systems.

More exciting applications are in real-estate portfolio management and corporate asset management (Castle 1993b). Portfolio managers typically manage large portfolios of real estate for major institutional investors such as pension funds. These managers analyse the entire nation's real-estate markets, deciding when to buy and when to dispose of properties. The data that are used to support these decisions are primarily geographical. GIS are used to help manage and integrate the data needed to make informed decisions about real-estate investment. Corporate users are much the same in that they have enormous real estate portfolios. For the most part, these real-estate portfolios consist of the property that a corporation uses to carry out its business, such as office and manufacturing space. They use GIS to analyse their holdings, deciding, for example, when to buy as opposed to rent (Castle 1993b).

11.6 Concluding comments

While this brief catalogue of applications is hardly exhaustive, the reader may notice that a wide variety of functions are enhanced using geographical technology, regardless of industry. Marketing, micro-marketing, site selection, cannibalization and competitive analysis, performance prediction, trade area analysis and network management all have strong geographical components that are improved by using business geographics.

12

Business, geography and academia in the UK

Graham Clarke, Paul Longley and Ian Masser

12.1 Introduction

In the previous chapter Nora Sherwood outlined the difficulties academic geographers have faced in the US in making meaningful links with the business world. She identified the particular problem that geography has never played a significant role in the academic upbringing of the 'captains of industry' and that geographers have also had a limited role in the private sector's rapid movement towards GIS. Hence, she concluded that there is generally no strong connection between academic geography and business in the United States. Despite her upbeat comments about the state of UK and European geography, some would argue that the situation in the UK is actually not appreciably different. Unwin (1992) suggests that the general public in the UK have a narrow perception of a geography education: 'Geography in Britain is thus still widely seen as being about "capes and bays" ' (p. 3); and 'In the political sphere geographers have played a very limited role in influencing decision making and in advising governments at national, let alone international, levels' (p. 3).

In the business world, Unwin (1986; 1992) points out that 50 per cent of employers surveyed in 1986 saw geography as a degree offering no particular advantages, and those that did see advantages pointed to computing and statistical attributes rather than any distinct geographical skills. Fortunately the picture does seem to be slowly changing. Geography does have a stronger tradition as an advanced level examination subject (16–18 age group) in the UK than in the US, and thus the younger element of the business sector may recognize contemporary geography as something more than 'capes and bays'. This is also generally the case with western European nations, particularly those which have a history of colonialism. However, it is likely to be a few years yet before this younger generation, who have had experience of quantitative geographical analysis, get their hands on the purse strings. In the UK, another potential caveat is the recent downgrading of geography (after age 13) as a core element of the National Curriculum in schools. This may ultimately reduce the demand for university places and, in turn, downgrade the importance of the discipline in the social sciences.

On a brighter note, geography in the UK (and generally across Europe) has a stronger traditional tie with the broad discipline of 'planning' than is the case in the US. These links exist either through career paths of geographers entering the planning profession or through membership within organizations such as the Association of Geographic Information, the Royal Town Planning Institute and the Institute of British Geographers. This has resulted in stronger links between academics, local authority researchers and planning consultants, and a tradition of policy-related research geared to service planning and resource allocation (Martin *et al.* 1992; Hirschfield *et al.* 1991; Worrall and Rao 1991; and Birkin *et al.* 1990). This at least has created some tradition of geographers collaborating with outside agencies. We might also take some comfort from the success of certain individuals in promoting geographical analysis (of whatever form) to the wider business sector. They have undoubtedly been helped by the relative success of geodemographics (see Batey and Brown, Chapter 5, this volume; Birkin, Chapter 6, this volume), which have whetted the appetites of many retail and service organizations for even greater analytical power. There are a number of geographers who have risen to senior management roles in organizations that use geographical information in some direct way (e.g. David Maguire, Paul Cresswell and Martin Clarke amongst the contributors to this volume) and also in a number of organizations that have been interested in analysing geographical information in-house (e.g., the in-house store-location units of major UK retail corporations such as Tesco, Sainsbury and Safeway are regular employers of geographers).

In this chapter we will explore some of the broader ways in which interactions between academia and the business and service sectors have been fostered in the UK. We will also suggest some possible lessons from this experience, and speculate on the prospects for future interaction between businesses, service providers and the academic research base within geography.

12.2 Geography and business links: case studies

Here we will begin to explore the nature of the interaction between academia and business through a number of different case studies. In each instance we will describe how this interaction has been fostered. In addition, we will speculate from the academic standpoint on some of the benefits and drawbacks of working with business organizations. More detailed analysis of specific methodologies and outcomes of individual projects have appeared in earlier chapters, and hence the emphasis here is not upon the work itself, but rather is upon the procedures and processes through which relationships between academia and business have been forged in the UK.

12.2.1 The Regional Research Laboratory Initiative

A key landmark in the development of GIS in the UK was the publication of the Chorley Report in 1987 (Department of the Environment 1987). Among its recommendations were the establishment of a centre for advice on the handling of geographical data in order to service the wider GIS community, and the encouragement of academics to 'actively disseminate information on the applications and benefits of GIS' (p. 122). At the same time, the Economic and Social Research Council (ESRC) was setting out a corporate strategy for data resource management. Given their new corporate aims, the ESRC responded to the Chorley Report by announcing the launch of the Regional Research Laboratory (RRL) Initiative in 1987.

The main objectives of the RRL Initiative were embodied in its title: the laboratory dimension indicated the need to establish a resource base for research and policy analysis; the research dimension highlighted the need for work on methodological issues relating to the management of large-scale data resources; and the regional dimension recognized the need to develop local centres of expertise capable of meeting the distinctive requirements of the different regional research communities within the UK. The trial phase of the RRL Initiative ran from February 1987 to October 1988 with the main phase beginning in October 1988 and continuing until December 1991. The sum allocated by ESRC for the main phase was over £2 million and eight RRLs were established covering the whole of the UK. The choice of RRL sites was based upon a peer-reviewed academic research competition, with the eight successful sites chosen from over 80 outline applications. The ESRC has since provided limited funds for the continued networking of the RRLs in order to sustain their common image, particularly in the international arena (Masser and Hobson 1992). A distributed network has been set up to manage these GIS research activities, participation in which has not been restricted to the RRLs (Walker 1993).

In terms of the interaction between academia and business the main RRL Initiative represented a fundamentally 'top-down' approach where the emphasis was very much upon fostering 'outreach' activity from academia to the wider community of (usually) public and private sector research. One of the main vehicles for developing such 'outreach' activities was the development of 'near market' research within the broad regional areas of the laboratories. It is pertinent to view these developments in the context of the political climate of the time, in that for much of the 1980s the *raison d'être* of social science research was in question, and the Initiative offered a convenient demonstration of the applied relevance of a social science subject such as geography. At the same time, traditional response mode research grants in the UK were being cut, thus reinforcing the importance to academia of attracting research funding directly from external private and public-sector institutions. There was some realignment of objectives and changes in performance measures over the period of the Initiative and here we will concentrate upon those pertinent to 'outreach' activity from academia. External income generation was reasonably consistently encouraged, and the RRLs were set the task of being self-supporting after the end of the Main Phase. The RRL Initiative can thus be seen in part as a four-and-a-half-year-long proactive attempt to foster 'near market' GIS research and outreach activity. The desired 'outreach' links could be viewed as:

ACADEMIA $\longrightarrow$ BUSINESS

The success of the RRL Initiative can be measured in a number of ways. Burrough and Boddington (1992) examine publications produced directly by RRL researchers as well as grant income received by the eight laboratories to give two more or less traditional 'academic' indicators of performance. A second measurement of the RRL success is the number and types of business links formed. Table 12.1 shows a representative selection of clients of the various RRLs (Economic and Social Research Council 1991). It illustrates the overwhelmingly public-sector orientation of the RRL client base, as well the role of leverage of additional funding from *within* the academic sector.

The less tangible output from the Initiative is the measurement of its promotional activities: that is, the raised awareness of the potential of GIS in a number of areas. This can be labelled 'soft consultancy'. We shall explore this theme under a number of headings. The most visible form of collaboration comes in the development of new hardware platforms and new software packages to run on these platforms. During the Initiative, RRL Scotland was given a considerable boost by a major research grant from Digital Equipment Corporation Europe to establish a parallel processing laboratory. The experience built up using this equipment enabled the RRL to make a successful bid, in conjunction with the Edinburgh Parallel Computing Centre, to the Department of Trade and Industry and the Science and Engineering Research Council for the development of parallel processing methods in GIS. This project has a number of international industrial partners both on the hardware (Meiko and DEC) and software (ESRI-UK, Laserscan and SmallWorld amongst others) sides (Healey *et al.* 1993). On the software front, important developments were ORCID (On-line Retrieval of Cartographic Data – plus an extra 'I'!), developed within the Wales and South West Regional Research Laboratory to enable users to browse map data on a graphics screen and select subsets of cartographic data to be written out to external software. More usually, however, 'collaboration' has entailed the wider adoption and use of proprietary GIS packages at the RRL sites (ARC/INFO in particular being made available through a nationwide licensing arrangement by the end of the Initiative), and a number of 'near market' research projects which have been concerned with the more routine use of GIS to perform operations such as overlay, buffering, etc. In this sense, the RRLs were consumers rather than suppliers of consultancy services. A second major focus of collaborative ventures with public-sector agencies was the access that this gave in many cases to data-sets held by these agencies. A recurrent theme

Table 12.1 A representative selection of clients of the UK Regional Research Laboratories

Avon Family Practitioner Committee
Blackpool, Wyre and Fylde Health Authority
British Council
British Rail
Castine Research Corporation
CDMS Ltd
City of Duisburg
Council of Europe
Department of the Environment (England and Northern Ireland)
Department of Health and Social Security (Northern Ireland)
Economic and Social Research Council
English Estates
ESRI
European Social Fund
Housing Corporation
IBM
Joseph Rowntree Foundation
Kenya Government
Leicester City Council
Leicestershire Constabulary
Leicestershire District Health Authority
Leverhulme Trust
Littlewoods Organization
Merseyside Information Service
Natural Environment Research Council
Nature Conservancy Council
North Tyneside Family Practitioner Committee
Northern Regional Health Authority
Nuffield Foundation
Office of Population Censuses and Surveys
Royal Statistical Society
Rural Development Commission
St Helens MBC/Merseyside Task Force
Science and Engineering Research Council
Scottish Development Department
Scottish Homes
Scottish Office (general)
Southmead District Health Authority
South Tyneside Metropolitan Borough Council
Strand studentship
Transport and Road Research Laboratory
Training Agency
Universiti Technologi Malaysia
University research committees (various)
Welsh Office

Source: ESRC (1991).

through this volume has been the inaccessibility of geographical information in disaggregate form, and direct interaction with the organizations which generate data has in certain circumstances led to concessions regarding the release of individual, or at least geographically disaggregate, information. In an era in which the level of social science research funding is not commensurate with the collection of large-scale primary data-sets, this has provided an important means of developing the range and scope of geographical analysis problems that may be tackled through empirical analysis – and in some ways represents an extension of previous research council initiatives which have sought to involve academics in the design and execution of large-scale public-sector surveys (e.g. the ESRC's 'Survey Link' scheme).

With regard to the RRL Initiative, this was particularly the case in the research on crime and policing that was carried out by the North-East RRL, and in work on epidemiology carried out by the North-East and North-West RRLs. With new data-sets many RRL researchers were able to combine GIS technology with their traditional social survey design skills. The Cardiff study outlined in Chapter 2 is also a good illustration of such a collaborative venture. In addition, the access to such new data sources provided by externally funded research has led, in turn, to a number of opportunities for RRL researchers to apply for traditional funding from university grant authorities.

The training element of staff at the RRLs has also been important. In the long term the creation of a core of highly qualified staff as a result of the Initiative is likely to have a lasting effect on the development of GIS research in the UK. One of the most distinctive features of the movement of RRL staff both during and after the Main Phase of the Initiative is the extent to which conventional boundaries between academia and the public and private sectors have been eroded at all levels (Masser and Hobson 1993). At the highest levels the movement of David Rhind from the South-East RRL to take up the post of Director General and Chief Executive of the Ordnance Survey represents a marked departure from conventional practices, as does David Maguire's move from the Midlands RRL to become Technical Director of ESRI (UK). However, movement into the public and private sectors has not been restricted to high-level appointments such as these. Masser and Hobson's study shows that more than a third of the RRL contract staff who have moved have gone to non-university posts in the public and private sectors. It is important to note that the study explicitly excludes Masters and undergraduate-level students who have also received training under individual RRLs.

The sudden cessation of funding at the end of the Main Phase of the Initiative has, however, left the future uncertain. Few, if any, of the RRLs have maintained the scale of operations that characterized their Main Phase activities, and some may in truth become little more than 'shell' operations. As a short-term fillip to the dissemination of GIS awareness and the establishment of universities in 'quasi-consultancy' roles, the RRL Initiative has undoubtedly been a success, but it may be questionable to suggest that this is the most cost-effective means of achieving a sustained and academically worthy presence in the applied research market.

12.2.2 GMAP (Geographical Modelling and Planning) Limited

The School of Geography at the University of Leeds has long been associated with quantitative geography and under the leadership of Alan Wilson has established a reputation in the fields of urban and regional modelling. Much of this work throughout the 1970s and 1980s was theoretical in nature and was used to illustrate the complexities of urban spatial structures (Wilson 1981; Clarke and Wilson 1985). The mid-1980s, however, saw a major shift in emphasis at Leeds from theory (alone) to applications (some of which were driven by new theoretical developments). This work began in the area of health-care modelling in West Yorkshire and in Piedmont in Italy. In particular, work began on making the outputs of models easier to understand and interpret for those planners (public and private sector) who remain unaccustomed to the terminology of spatial modelling.

The arrival of GIS was seen as producing a new opportunity for the development of such applied spatial modelling. It quickly became apparent to the Leeds team that GIS and spatial modelling were developing in parallel rather than in tandem: and, moreover, that if the benefits of both could be harnessed, then there existed the potential for powerful new products for the marketplace (cf. Birkin *et al.* 1987). Unlike the RRL initiative, the School of Geography at Leeds had no formal mission statements or requirements in terms of marketing its products beyond academia. Rather, a small but developing client base was built up based upon personal contacts and a development of a network of clients whose problems could be solved using the particular spatial-analysis skills developed at Leeds. If the RRL Initiative represented a coordinated 'top-down' approach to the interaction between academia and business then the story described in this section represents a 'bottom-up' response of one university department to the demand for applied consultancy within particular niches of the fast developing GIS research area. Formally, the interaction process could be represented as:

BUSINESS $\longrightarrow$ ACADEMIA

Table 12.2 illustrates the breadth of applications now undertaken at the University of Leeds, and it is interesting to compare this with Table 12.1. From humble origins (we can well remember the design of regression models to calculate average farm-path lengths for the UK Post Office, and the estimation of turnover for a potential new dry ski-slope in Pudsey, near Leeds!) the variety of retail and commercial sponsors has blossomed in an incremental fashion. Such has been the interest from the private sector that GMAP (Geographical Modelling and Planning) Limited was founded in the University of Leeds School of Geography in 1989. In 1994 GMAP employed 75 full-time staff, ranging from graduate geographers (crucial for maintaining spatial analysis and modelling skills), computer programmers (crucial for graphics and data-base management) and clerical staff. The Unit has since expanded into the American market with an exclusive deal with Polk Ltd (an agency which is involved with spatial data handling in the US) and into Australia through

Table 12.2 A representative selection of clients of GMAP Ltd

Retailing	W.H. Smith
	Storehouse
	Thorn EMI
	Asda
	Sainsbury (Homebase)
Motor industry	Toyota (UK, Belgium, US, Canada)
	Ford (UK, Europe)
	Polk
	Midas Muffler
Leisure	Whitbread
	Pizza Hut
Financial services	Bank of Scotland
	Leeds Permanent Building Society
	State Bank of South Australia

collaboration with the Applied Population Research Unit within the Department of Geography at the University of Queensland.

As the scale of GMAP itself has grown, so too has the geographical scale of its projects. From a pilot study of Oxfordshire and Merseyside in the UK in 1986, work with Toyota has subsequently expanded to the rest of the UK, Belgium and now Canada. The partnership with Ford began with the UK and has since been rolled out to all the remaining countries of western Europe.

A key question is: what conditions have promoted such a bottom-up approach to the business links? The most important factor is undoubtedly the focus at Leeds on spatial analysis and 'what if' planning coupled with the modern graphics environment standard to most GIS. The scope of operations has not broadened at a rate commensurate with the broadening of the GIS field, with much of GMAP's business concerned with the creation of customized systems, which often feature specialized spatial-analysis capabilities. In this sense at least, GMAP remains a specialized niche player in the global GIS market, emphasizing a level of spatial analysis and consultancy genuinely believed to be above that on offer through proprietary GIS packages. That is to say, many standard GIS packages are too inflexible for complex geographical problems, and *collaboration* is necessary. In this way, the client learns what is possible and feasible in the world of spatial analysis and the academics develop a greater understanding of business needs. The importance of this kind of relationship cannot be stressed strongly enough. Often the business user does not know at the outset exactly what is possible or actually most beneficial. It is important therefore that the design process is a two-way interaction between academic and user. Often the final version of a project in GMAP varies substantially from the pilot study first initiated, but is a far superior (and unique) product in the eyes of the client. Part of the reason for this is that in many cases, the institutions do not appreciate the value of the data which they hold.

Of course, most GIS packages are able to handle complex data structures and to transform data into maps, tables, graphs and such-like. However, after clients have become accustomed to pretty maps, many users then also want systems which can adequately address different business problems, problems which often involve 'what if' questions related to potential future growth. Such strategies cry out for the best of the spatial modelling techniques on which quantitative geographers have spent much of their time working (often in theoretical rather than application terms). Such modelling capability adds a new proactive role to GIS, enabling clients to reconsider market positions in the light of experimental 'what if' scenarios. One of the most flattering comments about a GMAP system came from a senior director of a major British retail company. Preliminary studies had revealed a national market penetration for this company which was very dense in the south-east, and which became progressively lower in the north of the country, with a two- to three-fold variation between regions. Similar kinds of intra-regional variation could be superimposed on this pattern. The director's comment was simply that it was eye-opening for the company to be able to see such obvious potential for further expansion, by building up in certain areas to market shares known to be achievable elsewhere. His board had grown so used to being told by analysts that it was a mature business which should concentrate on 'consolidation', that the organization had actually begun to believe it. One important principle which follows on from this can be stated roughly as follows: the benefit of an information system to a client is inversely proportional to the skill required to use it. Although GMAP systems are constructed using relatively sophisticated ideas, they are always presented as customized, menu-driven systems which are usable by complete computer novices.

There are undoubtedly pitfalls in such an approach. This is particularly so as the modelled outputs are not always unambiguous and need to be interpreted with care. What is provided is very much a decision-support system (Densham 1991), and not an automated replacement for the decision-making process. The key to the success of this process is again the establishment of the long-term relationship between client and consultant, involving mutual education. The consultant seeks better to understand the characteristics of the client's business environment. The client seeks to understand the underlying principles of the modelling approach, but without needing to dwell on technical minutiae. In order to be effective, the product as a whole is simply too complex for clients to be allowed access to the system code, and responsibility for all maintenance and updating of the system must remain with the consultant.

12.2.3 Teaching Company Schemes

The Teaching Company Scheme was launched in the UK in 1975 by the Science and Engineering Research Council and the Department of Trade and

Industry in order to help companies form partnerships with higher-education institutions. The stated objective of such partnerships was to improve company effectiveness through the development of new techniques and procedures. It is probably fair to say that, to date, applications have been dominated by manufacturing and engineering partnerships rather than by the business and commercial sectors which now dominate the UK economy. However, there are signs that this bias is shifting, and in this section we report on two examples of Teaching Company Scheme involvement with GIS.

The first has been set up between URPE-RRL-Liverpool with Credit and Data Marketing Services Ltd and builds upon links that have been built up since 1978 between the company and the Department of Civic Design at the University of Liverpool. Much of this collaboration has been directed towards the development of a location-assessment package for companies wishing to audit their locational efficiency. The Teaching Company Scheme is designed to offer academics who can bring to industry 'a significant and continuing research base' the opportunity to develop their knowledge and skills in an industrial or business environment. From the business organization side, firms have to be seen to be willing to demonstrate a 'commitment to change' based on the development work with academia. Key benefits for the research team at Liverpool RRL include the creation of research opportunities which extended the application of existing methodologies to new fields, as well as enhancing the Department's teaching programme and adding to the university's reputation for building links with business. Key benefits for the industrial partner include access to the technical expertise of the research team, and the marketing advantages associated with the academic standing of the team.

The second example is the partnership between the Midlands RRL and Dixons Stores Group Ltd (a UK town-centre electrical goods retailer). The aim is to build information systems which will 'significantly improve the quality and flow of information to headquarters managers, managers in retail outlets and to sales assistants' (Strachan 1994). The information base concerns both finance and sales information as well as checks on stock losses in distribution. The respective benefits to both parties are similar to the Liverpool example. Dixons do not currently possess suitably qualified staff to undertake the construction of these types of information system and thus benefit from the academic expertise in the Midlands RRL. In turn, the RRL gets the usual financial rewards from the grant funds to employ staff in new research areas, and access to the real data and applied problems of modern business.

Such partnerships could be described as examples of horizontal integration of business and academia:

BUSINESS ⟷ ACADEMIA

More generally, links in education and training were also developed within the 'top down' strategy of the RRL Initiative, for example through the secondment of staff and the development of job-sharing arrangements with private companies. Good examples of the former include the link between GENASYS and URPE-RRL-Manchester, while one of the best instances of the latter is the arrangement made between the South-East RRL and Small World.

12.3 Conclusions

The three case studies have attempted to illustrate some of the different ways in which it is possible to develop links between academia and business. Each has been associated with a different level of integration (top-down, bottom-up and horizontal, respectively). There are a number of themes which emerge from these case studies.

The first set of issues relate to commercial success and project durability. By most academic standards the RRL initiative has been a great success, generating 580 publications in total (Burrough and Boddington 1992). However, in most cases RRLs have been unable to maintain a level of income generation from externally funded projects to sustain a critical mass of key workers. By contrast, the GMAP model appears to be sustainable commercially. There is irony here, not least in that the formation of GMAP Ltd postdated unsuccessful applications by the University of Leeds School of Geography to participate in the Main Phase of the RRL Initiative. Whatever the strengths of GMAP in its present form, these apparently went unrecognized in the research bidding process, suggesting in retrospect an apparent failure of the mechanisms of academic peer review to anticipate 'near market' research potential – or perhaps merely that preoccupation with 'near market' research problems itself saps the time that it now takes to produce watertight research proposals. Perhaps the time span of the RRL Initiative was simply too short, with academic-sector funding being withdrawn, and with fast-changing academic fashions and limited resources preempting the provision of sufficient 'exit' funding.

And yet the relative success of GMAP clearly demonstrates that research council funding is not a necessary precursor to success. In part it must reflect product niche and the tailoring of solutions around business and service needs: yet it may also reflect the level of business enterprise shown by the academic partner. To service the needs of a client such as Ford Europe requires a level of investment in resources that few academics may be prepared to risk. If they do, then they must invest in marketing expertise in order to sustain orders and generate new business. At some stage there is a bifurcation point between academics 'doing some consultancy' (with or without indirect or direct short-term research council backing) and commitment to a real business which has to survive on its own merits. How many academics are also good at marketing, personnel management and ultimately senior business management? If they are not prepared at least to broach this question then it is unlikely they are the right ones to continue to build on the foundations already in place.

Whatever the scale of the interaction between academia and business it is clear that there has already been an enormous educational role to the development of such links. The RRLs have provided an important training ground for a number of researchers now active outside academia. If commercial pressures continue to generate demand for spatial skills then this is good news for the geography (and related) departments that have taken part in these collaborative projects. It must also be good news for the teaching programmes within these departments. Nothing improves the image of GIS and spatial-analysis-type courses more than examples and illustrations taken from real-world applications. Students are showing a renewed interest in computer-based and analytical courses as they see the perceived benefits in terms of new types of job opportunities. That has to be good news for GIS and quantitative geography.

And yet the 'success' of the RRL Initiative and of GMAP must also be qualified, particularly in so far as a central indicator of success is the number of quantitative geographers leaving academia to work in the business and service sectors. GMAP has sapped the University of Leeds School of Geography of most of the services of at least two highly respected academics, while there has been a more general criticism of the RRL Initiative that 'near market' research has led to an over-preoccupation with 'using other people's software to tackle other people's problems' at the expense of a focus upon methodological and substantive issues of intrinsic academic merit. Time will tell whether flirtation with the business sector is of sustained benefit for the continual regeneration of the research base, and how many of the 580 papers of the RRL Initiative will come to be regarded as scholarly citation classics. Some haemorrhaging of talent has been an inevitable consequence of broader UK trends in education policy in the 1980s and 1990s. However, the (at least partial) loss to quantitative geography of some of its most respected protagonists is potentially alarming, coming at a time at which academic fashions continue to deskill geography with respect to the elements of the curriculum which Unwin (1992) recognizes as central to employers' (e.g. business and service) needs.

A further consideration is that the attitudes of the academic establishment may be slower to change than the wider role of education and research in society. We have already commented upon the failure of academic peer review to identify 'winners' in the proposals for the RRL Initiative. Similar peer review in the UK forms the basis of the UK University Funding Council's (UFC's) periodic research selectivity exercises. Two comments might also be made about the most recent (1992) UFC rankings. First, despite securing large ESRC grants and numerous 'spin off' contracts (see Table 12.1), half of the main host departments of the RRL sites were rated no better than 'average' (rank 3) in the UFC league table of excellence. There is thus little evidence that 'near market' research and 'outreach' activity were associated with excellence by the review panels. Secondly, the particular reward for the rapid development of GMAP Ltd in the Leeds School of Geography appears to have been relegation from the top-ranked category achieved in the 1987 assessment to the next highest (rank 4) quintile. Just as the acknowledged success of the RRL Initiative was rewarded by failure to deliver further research council funding, so, in a peculiar British way, it may be necessary for university departments to steel themselves to apparent academic 'failure' as a prerequisite to commercial success in applied research.

What of the future? We have outlined three possible routes to building partnerships with business and commerce. The good news is that there is plenty of room left for newcomers, and geographers will quickly find that their skills *are* in demand. That can be both comforting and rewarding. It is primarily a question of willingness and ability. Time will tell whether these qualities will give quantitative geography and GIS continued relevance in the continual restructuring and change in the geographical landscapes of business and service provision.

Appendices

Appendix A   CERCO technical contacts

Country	Technical contact	Address	Phone and fax nos.
Austria	F. Hrbek	Austrian Federal Office of Metrology and Surveying Schiffamtsgasse 1—3 A 1025 Vienna	(43) 1 21176 ext. 3602 (43) 1 216106 2
Belgium	C. van Loon	Institut Géographique National Abbaye de la Cambre 13 1050 Brussels	(32) 2 648 64 80 (32) 2 646 25 42
Croatia	N. Francula	Geodetski Fakultet Kaciceva 26 Zagreb	(38) 49 442600 (38) 49 519305
Cyprus	N. Roussos	Dept of Lands and Surveys 29 Michalakopoulo Street Nicosia	(357) 2 302929 (357) 2 366171
Denmark	Peter Jacobsen	Kort-og Matrikelstyrelsen Reutemestervej 8 DK 2400 Copenhagen	(45) 3587 5050 (45) 3587 5059
Estonia	Mart Pork	Estonian State Land Department Mustamae Str 51 PO Box 1635 Talliuv EE0006	(372) 142 528202 (372) 142 528401
Finland	R. Nuuros	National Board of Survey PO Box 84 SF – 00521 Helsinki	(358) 0 154 5003 (385) 0 154 5005
France	B. Marty	Institut Géographique National Av. Pasteur 2 94160 Saint-Mande	(33) 1 439 88273 (33) 1 439 88445

Germany	Ingo Wilski	IFAG	(49) 69 633 33
		Richard-Strauss Allee 11	71
		D – 6000	
		Frankfurt A.M.	(49) 69 633 34 25
Greece	I. Spandideas	Hellenic Military Geographical	(30) 1 884281216
		Service	(30) 1 8817376
		4 Evelpidon Str	
		11362	
		Pedion Areos	
		Athens	
Hungary	Dr C. Remstey	Dept of Lands and Mapping	(36) 1 131 4130
	Fulopp	Ministry of Agriculture	(36) 1 111 2021
		H-1860 Budapest 55	or (36) 1 153
		PO Box 1	0518
Iceland	T. Bragason	Landmelingar Islands	(345) 1 681611
		(Iceland Geodetic Survey)	(354) 1 680614
		Laugavegi 178 Box 5060	
		125 Reykjavik	
Ireland	P. Prendercast	Ordnance Survey	(353) 1 8206100
		Phoenix Park	ext. 351
		Dublin 8	(353) 1 8204156
Italy	N. Donait	Capo Sazione	(39) 55 12775 –
		Relazioni Internazionali	267
		Istituto Geografico	(39) 55 282172
		Militara Italiano	
		Via Cesare	
		Battisti 10	
		50100 – Firenze	
Latvia	A. Lekuzis	Dept of Geodesy and	
		Cartography	(371) 0132
		11 Novembra	211263
		krasamala 31	(371) 0123
		IV-1484 Riga	225039
Lithuania	Rimas Eugenijus	State Department of Surveying	
	Paravinskas	and Mapping	(370) 0122 62 71
		A Jaksto 9	05
		2600 Vilnius	(370) 0122 62 76
			18
Luxembourg	André Majerus	Administration du Cadastre et	
		de la Topographie	(352) 449 01266
		BP 1761	(352) 449 01333
		L – 1017	
		Luxembourg	
Netherlands	P. W. Grudeke	Topografische dienst	(31) 5910 96201
		PO Box 115	(31) 5910 96296
		7800 AC Emmen	
Norway	J. M. Larsen	Statens Kartverk	(47) 67 19000
		N – 3300 Honefoss	(47) 67 18001
Poland	K. Firwitz	Ministry of Physical Planning	
		and Construction	(48) 2 628 73 64
		Department of Geodesy	(48) 2 628 58 87
		Cartography and Land	
		Management	
		Wspolna 2	
		PL – 00926 Warsaw	

Portugal	J. M. Barreiro Guedes	Instituto Geographico E Cadastral	(351) 1 609925
		Prays Da Estrela	(351) 1 3970248
		1200 Lisboa	
Slovak Republic	E. Ondrejicka	Geodesy, Cartography and Cadastre Authority	(42) 7 492002
		Hlboke Road 2	ext 5327
		813 23 Bratislava	(42) 7 497573
Slovenia	Bozena LIPEJ	Republiska geodetska uprava	(386) 1 312315
		Kristanova 1	(386) 1 122021
		61000 Ljubljana	
Spain	J. Rodriguez Sanchez	Centro Nacional de Información Geográfica	(34) 1 554 16 45
		C/General Ibanez de Ibero 3	(34) 1 553 29 13
		28003 Madrid	
Sweden	L. Ottoson	National Land Survey of Sweden	(46) 26 153423
		S - 801 82 Gavle	(46) 26 653160
Switzerland	F. Jeanrichard	Office Fédéral de Topographie Seftigenstrasse 264	(41) 31 9632111
		CH - 3084 Wabern	(41) 31 9632459
Turkey	O. Demirkol	General Command of Mapping	(90) 4 3197740
		TR 06100 Cebeci	ext. 225
		Ankara	(90) 4 3201495
UK (Great Britain)	N. Smith	Ordnance Survey	(44) 1703 792 052
		Romsey Road	(44) 1703 792 660
		Maybush	
		Southampton	
		SO9 4DH	
UK (Northern Ireland)	M. Brand	Ordnance Survey of Northern Ireland	(44) 1232 661244
		Colby House	(44) 1232 683211
		Stranmillis Court	
		Belfast BT9 5BJ	

Source: personal communication between R. Waters and N. Smith, Ordnance Survey Great Britain

Appendix B   US data providers

(NB **Source** identifies address and then telephone number — international prefix + 1)
Title: *American Housing Surveys, National Core and Supplement Files 1985, 1987 and 1989 and MSAs 1988 and 1989 Core Files*
Source: US Census Bureau Washington, DC 20233-8300 (301) 763-4100
Data media: CD-ROM
Key data: Housing, property value, neighbourhood quality, mortgage, rent, insurance; 48 000 housing unit sample includes year built, occupancy status, unit characteristics, monthly payments, utility costs, property insurance costs, rubbish collection costs, real estate taxes, government assistance, and neighbourhood quality measures
Data time span: 1985 to 1989
Geographic scale: United States and regions

Title: *Census of Agriculture Geographic Area Series – State and County Data (AC87-A)*
Source: US Census Bureau Washington, DC 20233-8300 (301) 763-4100
Data media: CD-ROM
Key data: Agriculture, farms: data on number of farms, land use, irrigation, crops, livestock, poultry, value of farm products sold, farms classified by specified characteristics, expenses, and operator characteristics. Size of farm, tenure of operator, type of organization, market value of agricultural products sold, age and principal occupation of operator, and industry grouping
Data time span: 1982 and 1987, with some comparative data for earlier years
Geographic scale: United States, states, counties, and outlying areas including Puerto Rico, Guam, Virgin Islands, American Samoa, and Northern Mariana Islands

Title: *Census of Manufacturers Geographic Area Series (MC87-A)*
Source: US Census Bureau Washington, DC 20233-8300 (301) 763-4100
Data media: Series of 51 paper-bound reports, one for each state and the District of Columbia
Key data: Industry; employment; payroll for industry groups; number of establishments; value added by manufacture; cost of materials consumed; and capital expenditures
Data time span: 1987, with some comparative data for 1982
Geographic scale: United States, states, metropolitan statistical areas (SMAs), counties, and selected places

Title: *Census of Manufacturers Industry Series File (MC87-1)*
Source: US Census Bureau Washington, DC 20233-8300 (301) 763-4100
Data media: CD-ROM
Key data: Industry; manufacturing; employment; payroll for three-digit SIC industry groups; quantity and value of products shipped and materials consumed; cost of fuels and electric energy consumed; capital expenditures; assets; rents; inventories; employment; payrolls; payroll supplements; hours worked; value added by manufacturing; number of establishments; and number of companies.
Data time span: 1987, with some comparative data for earlier years
Geographic scale: United States and states

Title: *Census of Population and Housing Summary Tape File 1A (STF-1A)*
Source: US Census Bureau Washington, DC 20233-8300 (301) 763-4100
Data media: 17 CD-ROMs
Key data: One hundred percent data for population and housing subjects. Population items tabulated include age, race, sex, marital status, Hispanic origin, household type, and household relationship. These subjects are cross-tabulated by age, race, sex, or Hispanic origin. Housing items tabulated include occupancy-vacancy status, tenure, units in structure, value, contract rent, meals included in rent, and number of rooms in housing unit
Data time span: 1990
Geographic scale: States; counties; county subdivisions; places and place segments, totals for split places, and remainder of county subdivision; census tracts/block numbering areas (BNAs), census tract/BNA segments, and totals for split census tracts/ BNAs; block groups (BC) and BG segments, and totals for split BG; American Indian and Alaska Native areas (AIANAs) or state portion of AIANAs; and congressional districts

Title: *Census of Population and Housing Summary Tape File 3A (STF-3A)*
Source: US Census Bureau Washington, DC 20233-8300 (301) 763-4100
Data media: 61 CD-ROMs

Key data: Population characteristics, ancestry, language, housing, education, occupation, income
Data time span: 1990
Geographic scale: States; metropolitan statistical areas (MSAs) or state portion of MSAs; urbanized areas (UAs) or state portion of UAs; counties; county subdivisions; places and place segments, totals for split places, and remainder of county subdivision; census tracts/block numbering areas (BNAs) and census tract/BNA segments, and totals for split census tracts/BNAs; block groups (BGs) and BG segments, and BG totals for split BGs, and American Indian and Alaska Native areas (AIANAs) or state portion of AIANA'S

Title: *Census of Retail Trade Geographic Area Series File (RC87-A)*
Source: US Census Bureau Washington, DC 20233-8300 (301) 763-4100
Data media: CD-ROM
Key data: Employment, sales, payroll, general statistics for establishments with payroll on number of establishments, sales, payroll, and employment
Data time span: 1987, with some comparative data for 1982
Geographic scale: United States, states, metropolitan statistical areas (MSAs), counties, and selected places

Title: *Census of Retail Trade Zipcode Statistics File (RC87-Z)*
Source: US Census Bureau Washington, DC 20233-8300 (301) 763-4100
Data media: CD-ROM
Key data: Employment, sales, payroll; number of retail establishments classified by size and kind of business by ZIP code
Data time span: 1987
Geographic scale: States and five-digit zipcodes

Title: *Census of Service Industries Geographic Area Series File (SC87–A)*
Source: US Census Bureau Washington, DC 20233-8300 (301) 763– 4100
Data media: CD-ROM
Key data: Health services, professional services, hotels and motels, automotive repair. Statistics cover hotels, motels, and other lodging places; personal and business services; automotive repair, services, and parking; miscellaneous repair services; and amusement and recreation services, including motion pictures. The report also includes health services; legal services; some educational services; social services; museums, art galleries, botanical gardens, and zoos
Data time span: 1987, with some comparative data for 1982
Geographic scale: United States, states, metropolitan statistical areas (MSAs), counties, and selected places

Title: *Census of Transportation Geographic Area Series File (TC87–A)*
Source: US Census Bureau Washington, DC 20233-8300 (301) 763-4100
Data media: CD-ROM
Key data: Transportation; employment; revenue; payroll for water transportation; transportation services (except by air); and motor freight transportation and warehousing
Data time span: 1987, with some comparative data for 1982
Geographic scale: United States, states, and selected metropolitan areas

Title: *Consumer Expenditure Surveys*
Source: US Department of Labor Bureau of Labor Statistics, Washington, DC 20212 (202) 523-1221 Contact John Rogers at (202) 606-6900
Data media: diskette (5.25 in or 3.5 in). **Data format:** ASCII or Lotus 1-2-3.

Key data: Consumer expenditures; income; average annual out-of-pocket expenditures for housing, apparel, services, transportation, health care, entertainment, personal care, reading, education, personal insurance, and pensions. Diary survey data include average weekly expenditures on frequently purchased items such as food at home, food away from home, alcoholic beverages, tobacco products and smoking supplies, personal care products and services, and nonprescription drugs and supplies
Data time span: 1984 to 1991
Geographic scale: United States, states, or local governmental areas, varying by file

Title: *County and City Data Book: 1988 Files*
Source: US Census Bureau Washington, DC 20233-8300 (301) 763-4100
Data media: Diskettes and CD-ROM
Key data: population; housing; income: employment; education; health care; climate
Data time span: 1980 to 1986
Geographic scale: Regions, divisions, states, counties, incorporated cities of 25 000 or more, and places of 2500 or more

Title: *County Statistics Tape File 4 (CO-STAT 4)*
Source: US Census Bureau Washington, DC 20233-8300 (301) 763-4100
Data media: Available on one tape reel at 6250 bpi
Key data: Business; construction; crime; education; government; health; housing; labour; income
Data time span: 1960 to 1988
Geographic scale: States and counties or county equivalents

Title: *Current Population Survey*
Source: US Census Bureau Washington, DC 20233-8300 (301) 763-4100
Data media: CD-ROM
Key data: Labour; employment; occupation by population characteristics (race, age. sex. marital status, education)
Data time span: 1988 to 1992 (March of each year)
Geographic scale: Regions, divisions, all states, 268 metropolitan statistical areas (MSAs) and 300 central cities in multi-central-city MSAs

Title: *Economic Census Summary – 1987*
Source: US Census Bureau Washington, DC 20233-8300 (301) 763-4100
Data media: CD-ROM
Key data: 1987 economic census data by SIC code: labour, employment, payroll, sales, and minority group
Data time span: 1987, with some comparative data for earlier years
Geographic scale: states, counties, and zipcodes

Source: Thrall and Smeish (1994)

Appendix C Geographical information service providers (UK unless otherwise stated)

CACI Limited, CACI House, Kensington Village, Avonmore Road, London W14 8TS.
CCN Marketing, Talbot House, Talbot Street, Nottingham NG1 5HF.
Claymore Services Limited, Station House, Whimple, Exeter, EX5 2QH.
CMT Limited, Teddington, Middlesex.
ERDAS, 2801 Buford Highway, Suite 300, Atlanta, GA 30329, USA.
ERDAS (UK) Limited, Sheraton House, Castle Park, Cambridge CB3 0AX.

ERSI, 380 New York Street, Redlands, CA 92373, USA.

ESRI (UK) Limited, 23 Woodford Road, Watford, Hertfordshire WD1 1PB.

EuroDirect Limited, 2 Baptist Place, Bradford, BD1 2PS.

Genasys II Limited, 3rd Floor, Axis 63, Cross Street, Sale, Cheshire. (GENASYS US tel. 1-800-447-0265)

GMAP Limited, GMAP House, Cromer Terrace, Leeds LS2 9JU.

IDRISI Project, Graduate School of Geography, Clark University, 950 Main Street, Worcester, MA 01610, USA.

Intergraph Corporation, Huntsville, AL 35894, USA.

Intergraph, Delta Business Park, Great Western Way, Swindon, Wiltshire SN5 7XP.

Laser-Scan Limited, Cambridge Science Park, Milton Road, Cambridge CB4 4FY.

NDL International, Port House, Plantation Wharfe, London SW11 3TY.

Pinpoint Analysis Limited, Tower House, Southampton Street, London WC2E 7HN.

Public Attitude Surveys (PAS) Limited, Rye Park House, London Road, High Wycombe, Buckinghamshire HP11 1EF.

Retail Locations, 30–31 The Broadway, Woodford, Essex IG8 0HQ.

Smallworld Systems Limited, Burleigh House, 13–15 Newmarket Road, Cambridge CB5 8EG.

SPA Marketing Systems, 1 Warwick Street, Leamington Spa, Warwickshire CV32 5LW.

Stats MR, Gloucester House, Smallbrook Queensway, Birmingham B5 4HP.

Tactics International, 3rd Floor, 16 Haverhill Street, Andover, MA 01810, USA.

Target Group Index, Produced by British Market Research Bureau, Saunders House, 53 The Mall, London W5 3TE.

The TMS Partnership, Oxford House, 182 Upper Richmond Road, London SW15 2SH.

Transcad, Caliper Corporation, 1172 Beacon Street, Newton, Massachusetts 02161, USA.

TYDAC Technologies, 2 Venture Road, Chilworth Research Centre, Southampton SO1 7NP.

Unisys, Stonebridge Park, London NW10 8LS.

Glossary of abbreviations and technical terms

ACORN 'A Classification of Residential Neighbourhoods': a geodemographic discriminator which has been used extensively to examine variation in consumer behaviour between area types (UK)

AI Artificial intelligence

BPI Buying Power Index (US)

CCJ County Court judgement (UK)

CPD Central postcode directory (UK)

DBMS database management system

DIME the Dual Independent Map Encoding system developed for the 1970 US Census, initially for metropolitan areas only

ED enumeration district: a census area for which a single census enumerator has responsibility (England, Scotland and Wales)

ER Electoral Register (UK)

ESRC Economic and Social Research Council (UK)

EFTPOS electronic fund transfer at point-of-sale

EPOS electronic point-of-sale

GA genetic algorithm

GISSAS combined 'Geographical Information System' and 'Spatial Analysis Software' (see Section 8.6.2)

GRO(S) General Register Office (Scotland)

MAUP Modifiable Areal Unit Problem: a problem inherent to much of geographical analysis, which arises because different types and levels of aggregation can produce wholly different representations of geographical phenomena

NRS National Readership Survey (UK)

OA Census Output Area (Scotland)

PAF the Postcode Address File, which provides a link between all addresses and a postcode, and also indicates the census enumeration districts within which these postcodes fall (UK)

PED pseudo-enumeration district (see Figure 4.3)

PPU part postcode unit (see Figure 4.3)

OPCS Office of Population Censuses and Surveys (England and Wales)

RDBMS Relational Database Management System

RRL Regional Research Laboratory (UK)

SAR	Sample of Anonymized Records: a 1% sample of households and a 2% sample of individuals from the 1991 UK Census, for which the complete anonymized records are available
SAS	Small Area Statistics: a set of counts providing information in around 9000 cells for each ED in England and Wales, or Output Area in Scotland
SDSS	Spatial Decision Support System (see Section 10.4)
SMMT	the UK Society of Motor Manufacturers and Traders
SMS	Special Migration Statistics: detailed information on flows of migrants, derived from the UK census
SWS	Special Workplace Statistics: detailed information on journeys-to-work, derived from the UK census
TGI	Target Group Index information collected about product purchases and media use, cross-tabulated against types (clusters) of census wards (UK)
TIGER	Topologically Integrated Geographical Coding and Referencing files produced to accompany the 1990 census. They are based upon lines which form both a description of the street network, and also a set of boundaries for the representation of census block bound-aries
UPRNs	Unique Property Reference Numbers (hierarchical customer references used by most utilities)
URISA	Urban and Regional Information Systems Association (US)
VAR	Value added reseller: an organization which 'adds value' to data. They generally are more knowledgeable about a specific industry, such as insurance or health care, than is a GIS vendor. VARs are developing application-specific menus and screens and providing customized training and support. Most common in the US, where cost recovery by public-sector data suppliers is low.

References

Abel D J, P J Kilby and J R Davis (1994) The systems integration problem, *International Journal of Geographical Information Systems* **8**(1); 1–12

Abler R F (1987) The National Science Foundation Center for Geographic Information and Analysis, *International Journal of Geographical Information Systems* **1**(4); 303–26

Abler R F (1993) Everything in its place: GPS, GIS, and geography in the 1990s, *The Professional Geographer* **45**(2); 131–9

Adams B F (1960) Geographic education in the public and parochial schools of a four-county sampling of Pennsylvania. Unpublished doctoral dissertation, Pennsylvania State University, Department of Geography, Pennsylvania

Allen S (1993) Levi Strauss & Company finds the perfect fit, *Business Geographics* **1**(4); 30–2

Allinson J (1993) The breaking of the third wave: the demise of GIS, *Planning Practice and Research* **8**(2); 30–3

Alper L and R Gelty (1969) Product positioning by behavioral life-styles, *Journal of Marketing* **33**

Anselin L (1989) What is special about spatial data? alternative perspectives on spatial data analysis. NCGIA Technical Paper 89–4, NCGIA, Santa Barbara, California

Applebaum W (1965) Can store location be a science? *Economic Geography* **41**; 234–7

Baker K (1989) Using geodemographics in market research surveys, *Journal of the Market Research Society* **31**; 37–44

Barr B (1993a) Census geography II: A review. In A Dale and C Marsh (eds) *The 1991 Census user's guide*, London: HMSO, pp. 70–83

Barr B (1993b) Mapping and spatial analysis. In A Dale and C Marsh (eds) *The 1991 Census user's guide*, London: HMSO, pp. 248–68

Barrow D (1993) The use and application of genetic algorithms, *Journal of Targeting, Measurement, and Analysis for Marketing* **2**; 30–4

Bateman M, K Jones and G Moon (1985) *Hampshire 1981 Census atlas*, Department of Geography, Portsmouth Polytechnic

Batey P W J and P J B Brown (1994) Design and construction of geodemographic targeting systems: what have we learnt in the last ten years?, *Journal of Targeting, Measurement and Analysis for Marketing* **23**; 105–115

Beaumont J R (1989) Market analysis, *Environment and Planning A* **21**; 567–9

Beaumont J R (1991) *An introduction to market analysis*: Concepts and Techniques in Modern Geography 53, Norwich Environmental Publications

Beaumont J R and K Inglis (1989) Geodemographics in practice: developments in Britain and Europe, *Environment and Planning A* **21**; 587–604

Bell W (1955) Economic, family and ethnic status: an empirical test, *American Sociological Review* **20**; 45–52

Berry B J L and F E Horton (1970) *Geographic perspectives on urban systems*, Englewood Cliffs, NJ: Prentice-Hall

Berry B J L and D F Marble (eds) (1968) *Spatial analysis: a reader in statistical geography*, Englewood Cliffs, NJ: Prentice-Hall

Berry J K (1987) Fundamental operations in computer-assisted map analysis, *International Journal of Geographical Information Systems* **1**(2); 119–36

Besag J and J Newell (1991) The detection of clusters in rare diseases, *Journal of the Royal Statistical Society A* **154**; 143–55

Birkin M (1994) The case of the missing performance indicators. In C S Bertuglia, G P Clarke and A G Wilson (eds) *Modelling the city*, London: Routledge, pp. 105–20

Birkin M and M Clarke (1988) SYNTHESIS: a synthetic spatial information system with methods and examples, *Environment and Planning A* **20**; 645–71

Birkin M and M Clarke (1989) The generation of individual and household incomes at the small area level using SYNTHESIS, *Regional Studies* **23**; 535–48

Birkin M, G P Clarke, M Clarke and A G Wilson (1987) Geographical information systems and model-based locational analysis: a case of ships in the night or the beginnings of a relationship? Working paper 498, School of Geography, University of Leeds

Birkin M, G P Clarke, M Clarke and A G Wilson (1990) Elements of a model-based GIS for the evaluation of urban policy. In L Worrall (ed.) *Geographical information systems: developments and applications*, London: Belhaven, pp. 133–62

Birkin M, G P Clarke, M Clarke and A G Wilson (1995) *Intelligent GIS*, Harlow: Longman

Black S and D J Cowen (1991) Information infrastructure for economic development: a unique state and university partnership for GIS implementation. Conference Proceedings of the Urban and Regional Information Systems Association

Boots B N (1986) *Voronoi (Thiessen) polygons*. Concepts and Techniques in Modern Geography 45, Norwich: Environmental Publications

Bracken I (1981) *Urban planning methods: research and policy analysis*, London: Methuen

Bracken I and D Martin (1989) The generation of spatial population distributions from census centroid data, *Environment and Planning A* **21**; 537–43

Broome F R and D B Meixler (1990) The TIGER database structure. In R W Marx (ed.) *The Census Bureau's TIGER system*, Bethesda MD: ACSM, pp. 39–47

Brown P J B (1988) A Super Profile based affluence ranking of OPCS urban areas, *Built Environment* **14**; 118–34

Brown P J B (1991) Exploring geodemographics. In I Masser and M J Blakemore (eds) *Handling geographical information*, London: Longman, pp. 221–58

Brown P J B and P W J Batey (1987a) A national classification of 1981 Census enumeration districts, the derivation of Super Profile Area Types. Area Classification Information Note 1, Centre for Urban Studies, University of Liverpool

Brown P J B and P W J Batey (1987b) The Super Profile lifestyle and target market national classifications: their derivation and description. Area Classification Information Note 3, Centre for Urban Studies, University of Liverpool

Brown P J B and P W J Batey (1987c) Lifestyle pen pictures, descriptions and comparisons of the lifestyle area types of the Super Profile Classification System. Area Classification Information Note 4, Centre for Urban Studies, University of Liverpool

Brown P J B and P W J Batey (1990) Geodemographics and the construction of individual-level market classifications: the case of the home-shopping industry. URPERRL Working Paper 16, Department of Civic Design, University of Liverpool

Brown P J B and P W J Batey (1994) The development of the 1991-based Super Profiles geodemographic classification system, URPERRL Working Paper 40, Department of Civic Design, University of Liverpool

Brown P J B, A F G Hirschfield and P W J Batey (1991) Applications of geodemographic methods in the analysis of health condition incidence data, *Papers in Regional Science: Journal of the Regional Science Association International* **70**(3); 329–44

Brusegard D and G Menger (1989) Real data and real problems: dealing with large spatial databases. In M Goodchild and S Gopal (eds) *Accuracy of spatial databases*, London: Taylor & Francis, pp. 177–85

Burkhart K (1993) Geographic analysis ensures customer satisfaction, *Business Geographics* **1**(1); 26–7

Burrough P A (1986) *Principles of geographical information systems for land resources assessment*, Monographs on Soil Resources Survey 12, Oxford: Oxford University Press

Burrough P A (1991) Soil information systems. In D J Maguire, M F Goodchild and D W Rhind (eds) *Geographical information systems: principles and applications*, London: Longman, pp. 153–69

Burrough and Boddington (1992) The UK Regional Research Laboratory Initiative, 1987–91, *International Journal of Geographical Systems* **6**; 425–40

Business Geographics (1993) Teachers sharpen geography teaching skills, *Business Graphics Briefcase* **1**(5); 14

Buxton R (1989) Integrated spatial information systems in local government – is there a financial justification?, *Mapping Awareness* **2**(6); 14–16

Buxton T (1993) Consumer packaged goods industry faced micro marketing challenge, *Business Geographics* **1**(5); 16

CACI (1993a) ACORN. Product brochure, CACI Information Services, London

CACI (1993b) Marketing Systems Today **8**(1), CACI Information Services, London

Carroll B (1992) Developments in the technology and practice of target marketing consumers in North America, *Journal of Targeting, Measurement and Analysis for Marketing* **1**; 107–12

Cathelat B (1993) *Socio-styles*, London: Kogan Page

Carver S (1991) Error modelling in GIS: who cares? In J. Cadoux-Hudson and I. Heywood (eds) *Geographic Information 1991*, Yearbook of the Association for Geographic Information, London: Taylor & Francis, pp. 229–234

Cassettari S (1993) *Introduction to integrated geo-information management*, London: Chapman & Hall

Castle G H (1993a) *Profiting from a geographic information system*, Fort Collins, CO: GIS World Inc

Castle G H (1993b) Location, location, location drives GIS in real estate, *Business Geographics* **1**(3); 30–2

Castle G H (1993c) GIS: a time-saving tool for targeting candidate properties, *Business Geographics* **1**(2); 20

CCN (1993a) EuroMOSAIC. Product brochure, Nottingham: CCN Marketing

References

CCN (1993b) MOSAIC: alternative approaches to classification, Nottingham: CCN Marketing

Charlton M, S Openshaw and C Wymer (1985) Some new classifications of census enumeration districts in Britain: a poor man's ACORN, *Journal of Economic and Social Measurement* **13**; 69–96

Chisholm M (1971) Geography and the question of 'relevance', *Area* **3**(2); 65–8

Clark A M and F G Thomas (1990) The geography of the 1991 Census, *Population Trends* **60**; 9–15

Clarke G P and A G Wilson (1994) Performance indicators in urban planning: the historical context. In C S Bertuglia, G P Clarke and A G Wilson (eds) *Modelling the city: performance, policy and planning*, London: Routledge, pp. 4–19

Clarke M and A G Wilson (1985) The dynamics of urban spatial structure, *Transactions, Institute of British Geographers* **10**; 427–51

Clifford D (1993) The Vauxhall dealer area analysis package, *Mapping Awareness* **7**(5); 25–7

Cooke D (1989) TIGER and the post-GIS era, *GIS World* July/August; 40–55

Crain I K and C L MacDonald (1984) From land inventory to land management, *Cartographica* **21**; 40–6

CRU/OPCS/GRO(S) (1980) *People in Britain: A census atlas*, London: HMSO

Cuzick J and J Edwards (1990) Tests for spatial clustering in heterogenous populations, *Journal of the Royal Statistical Society B* **52**; 73–104

Daedalus (1992) A new era in computing, *Journal of the American Academy of Arts and Sciences* Winter

Dale A (1993) The OPCS longitudinal study. In A Dale and C Marsh (eds) *The 1991 Census user's guide*, London: HMSO, pp. 312–29

Dale A and C Marsh (1993) *The 1991 Census user's guide*, London: HMSO

Dangermond J (1993) Overview of business applications for GIS, GIS in Business '93: Conference Proceedings, GIS World, Colorado, pp. 7–10

Daniel L (1993) Enhanced modeling helps position business geographics in the retail industry, *Business Geographics* **1**(5); 37–9

Davis P R and M A Schwartz (1993) ArcCAD applications to integrated surface and ground water model (ISGW) data preparation. Proceedings of Thirteenth Annual ESRI User Conference, vol. 3, Redlands: ESRI, pp. 341–50

Densham P (1991) Spatial decision support systems. In D J Maguire, M F Goodchild and D W Rhind (eds) (1991) *Geographical information systems: principles and applications*, London: Longman, pp. 403–12

Densham P J and G Rushton (1988) Decision support systems for locational planning. In R Golledge and H Timmermans (eds) *Behavioural modelling in geography and planning*, London: Croom Helm, pp. 56–90

Department of Education and Science, National Curriculum (1991) Geography in the National Curriculum (England). Leaflet, March, Department of Education and Science, London

Department of the Environment (1987) *Handling geographic information*. The report of the Committee of Enquiry chaired by Lord Chorley, London: HMSO

Diggle P J (1983) *Statistical analysis of spatial point patterns*, London: Academic Press

Ding Y and A S Fotheringham (1992) The integration of spatial analysis and GIS, *Computers, Environment and Urban Systems* **16**; 3–19

Dongarra J J, H W Meuer and E Strohmaier (1993) Top 500 Supercomputers, RUM 33/93, Computing Centre, University of Mannheim, Mannheim, Germany

Dugmore K (1992) 1991 Census: outputs and opportunities. In J Cadoux-Hudson and I Heywood (eds) *Geographic information 1992/3*, Yearbook of the Association for Geographic Information, London: Taylor and Francis, pp. 254–61

Economic and Social Research Council (1991) The Regional Research Laboratory Initiative: an interim evaluation, Swindon: ESRC

Elliott C (1991) Store planning with GIS. In J Cadoux-Hudson and D I Heywood (eds) *Geographic information 1991*, Yearbook of the Association for Geographic Information, London: Taylor and Francis, pp. 169–72

Evans N and R J Webber (1994) Advances in geodemographic classification techniques for target marketing, *Journal of Targeting, Measurement and Analysis for Marketing* **2**; 313–321

Fischer M M and P Nijkamp (eds) (1993) *Geographical information systems, spatial modelling and policy evaluation*, Berlin: Springer-Verlag

Flowerdew R (1991) Clustered Residential Area Profiles and beyond, *Mapping Awareness* **5**(3); 34–9

Flowerdew R and W Goldstein (1989) Geodemographics in practice: developments in North America, *Environment and Planning A* **21**; 605–16

Fotheringham A S and P A Rogerson (1993) GIS and spatial analytical problems, *International Journal of Geographical Information Systems* **7**(1); 3–19

Fotheringham A S and P A Rogerson (1994) *GIS and spatial analysis*, London: Taylor & Francis

Foust B and H Botts (1993) Mapping market penetration to discover 'hidden' opportunities, *Business Geographics* **1**(2); 38–40

Freeling J (1993) Using drive times to construct trading areas, GIS in Business '93 Conference Proceedings, pp. 261–4

Furness P (1992) Applying neural networks in database marketing: a review, *Journal of Targeting, Measurement, and Analysis for Marketing* **1**; 152–69

Gandy W E (1960) The status of geography in the public senior high schools in California. Dissertation Abstracts 20 4347, University Microfilms Order No. 60–1347, University Microfilms, Ann Arbor, MI, USA

Gatrell A C (1989) On the spatial representation and accuracy of address-based data in the United Kingdom, *International Journal of Geographical Information Systems* **3**(4); 335–48

GIS Strategies (1993a) **2**(1); GIS World Inc, Fort Collins, CO, USA

GIS Strategies (1993b) **2**(2); GIS World Inc, Fort Collins, CO, USA

GIS Strategies (1993c) **2**(3); GIS World Inc, Fort Collins, CO, USA

GIS World (1993) International GIS Sourcebook, GIS World Inc, Fort Collins, CO, USA

Gittus E (1963–64) An experiment in the identification of urban sub-areas, *Transactions of the Bartlett Society* **2**; 108–35

Gittus E (1964) The structure of urban areas: a new approach, *Town Planning Review* **35**; 5–20

Goodchild M F (1988) A spatial analytical perspective on geographical information systems, *International Journal of Geographical Information Systems* **1**; 327–34

Goodchild M F (1992) Geographical information science, *International Journal of Geographical Information Systems* **6**(1); 1–45

Goodchild M F, L Anselin and U Deichmann (1993) A framework for the areal interpolation of socioeconomic data, *Environment and Planning A* **25**; 383–97

Goodchild M F, R Haining and S Wise (1992) Integrating GIS and spatial data analysis: problems and possibilities, *International Journal of Geographical Information Systems* **6**(5); 407–23

Greater Manchester Council (1975) Multivariate analysis of the 1971 Census: a classification of wards in Greater Manchester. Discussion Note, 111, Greater Manchester County Planning Department, Manchester

Grimshaw D J (1994) *Bringing geographical information systems into business*, Harlow: Longman

Guy C M (1994) *The retail development process*, London: Routledge

Haining R (1990) *Spatial data analysis in the social and environmental sciences*, Cambridge: Cambridge University Press

Harvey D (1989) From models to Marx: notes on the project to 'remodel' contemporary geography. In W Macmillan (ed) *Remodelling Geography*, Oxford: Blackwell, pp. 211–16

Hawley A and O D Duncan (1957) Social area analysis: a critical appraisal, *Land Economics* **33**; 337–45

Healey R G, S Dowers, B M Gittings and T C Waugh (1993) GIS and parallel processing: research developments at RRL Scotland. In D Walker (ed) *Regional Research Laboratories Network 1993*, Department of Town and Regional Planning, University of Sheffield

Hirschfield A F G (1995) *Urban deprivation: problems, processes and counter measures in developed countries*, London: Belhaven (in press)

Hirschfield A F G, P J B Brown and J Marsden (1991) Database development for decision support and policy evaluation. In L Worrall (ed.), *Spatial analysis and spatial policy using geographical information systems*, London: Belhaven, pp. 152–87

Howard N (1969) Least squares classification and principal component analysis: a comparison. In M Dogan and S Rokkan (eds) *Quantitative ecological analysis in the social sciences*, Cambridge MA: MIT Press, pp. 397–412

Howe A (1991) Assessing potential of branch outlets using GIS. In J Cadoux-Hudson and D I Heywood (eds) *Geographic Information 1991*, Yearbook of the Association for Geographic Information, London: Taylor & Francis, pp. 173–5

Huff D L (1963) A probabilistic analysis of shopping center trade areas, *Land Economics* **39**; 81–90

Jackson B L (undated) SAS macros for converting between ARC/INFO Import/Export files and SAS data sets. Unpublished manuscript, Oak Ridge National Laboratory, Oak Ridge, USA

Jarman B (1983) The identification of underprivileged areas, *British Medical Journal* **286**; 1705–9

Jefferis D (1993) SpaAM: a spatial analysis and modelling system. Proceedings of Thirteenth Annual ESRI User Conference, vol. 3, Redlands: ESRI, pp. 79–87

Johnston R J, D Gregory and D Smith (eds) (1991) *The dictionary of human geography*, Oxford: Blackwell

Joint Working on Structure Plans in the Greater Manchester Area (1972) Multivariate analysis of the 1966 Census. Technical Working Paper 11, Lancashire County Planning Department, Preston

Kay W (1987) *Battle for the High Street*, London: Piatkus

Kehris E (1990a) A geographical modelling environment built around ARC/INFO. North West Regional Research Laboratory, Research Report 13, NWRRL, Lancaster

Kehris E (1990b) Spatial autocorrelation statistics in ARC/INFO. North West Regional Research Laboratory, Research Report 16, NWRRL, Lancaster

Kelly F (1969) Classification of urban areas, *Quarterly Bulletin of the Research and Intelligence Unit* **9**; 13–19, Greater London Council, London

Kelly F (1971) Classification of the London Boroughs. Greater London Research and Intelligence Unit Research Report 9, GLC, London

Keltecs (1989) The Cardiff House Condition Survey. Phase 1: Inner Area Final Report. Keltecs, Consulting Architects and Engineers Ltd, Grove House, Talbot Road, Talbot Green, Cardiff, UK

Kitagawa E M and K E Tacuber (1963) *Local community factbook: Chicago Metropolitan Area, 1960,* Chicago: Chicago Community Inventory, University of Chicago

Knox P (1978) Territorial social indicators and area profiles, some cautionary observations, *Town Planning Review* **49**; 75–83

Kubo S (1991) The development of GIS in Japan. In D J Maguire, M F Goodchild and D W Rhind (eds) *Geographical information systems: principles and applications,* London: Longman, pp. 47–56

Kunze E A (1993) GIS delivers logistics solutions to UPS, *Business Geographics* **1**(4); 34–7

Langford M, D J Maguire and D J Unwin (1991) The area interpolation problem: a Monte Carlo simulation. In I Masser and M Blakemore (eds) *Handling Geographical Information: Methodology and Potential Applications,* London: Longman, pp. 55–77

Laurini R and D Thompson (1992) *Fundamentals of spatial information systems,* London: Academic Press

Lein J K (1992) Modelling environmental impact using an expert-geographic information system, Proceedings of the GIS/LIS 1992 Annual Conference, pp. 436–44

Leventhal B, C Moy and J Harding (eds) (1993) *An introductory guide to the 1991 Census,* Henley: NTC Publications

Liverpool City Council (1969) Social malaise in Liverpool: interim report on social problems and their distribution, Liverpool: Liverpool City Planning Department

Lo C P (1989) A raster approach to population estimation using high-altitude aerial and space photographs, *Remote Sensing of Environment* **27**; 59–71

Maguire D J (1991) An overview and definition of GIS. In D J Maguire, M F Goodchild and D W Rhind (eds) *Geographical information systems: principles and applications,* London: Longman, pp. 9–20

Maguire D J (1992) The raster GIS design model – a profile of ERDAS, *Computers & Geosciences* **18**(4); 463–70

Maguire D J and J Dangermond (1991) The functionality of GIS. In D J Maguire, M F Goodchild and D W Rhind (eds) *Geographical information systems: principles and applications,* London: Longman, pp. 319–25

Maguire D J, M F Goodchild and D W Rhind (eds) (1991) *Geographical information systems: principles and applications,* London: Longman

Majure J J and N Cressie (1993) EXPLORE: exploratory spatial analysis in ARC/INFO. Proceedings of Thirteenth Annual ESRI User Conference, vol. 1, Redlands: ESRI, pp. 277–81

Manchester City Council (1973) Social information study of Manchester City: a brief outline of the scope and purpose of the study, Social Services Departent and City Planning Department, Manchester City Council, Manchester

Marsh C (1993a) Privacy, confidentiality and anonymity in the 1991 Census. In A Dale and C Marsh (eds) *The 1991 Census user's guide,* London: HMSO, pp. 111–28

Marsh C (1993b) The sample of anonymised records. In A Dale and C Marsh (eds) *The 1991 Census User's Guide,* London: HMSO, pp. 295–311

Martin D (1989) Mapping population data from zone centroid locations, *Transactions of the Institute of British Geographers* **14**; 90–7

Martin D (1991) *Geographic information systems and their socioeconomic applications*, London: Routledge

Martin D (1992) Postcodes and the 1991 Census of Population: issues, problems and prospects, *Transactions of the Institute of British Geographers* **17**; 350–7

Martin D (1993) *The 1991 UK Census of Population*. Concepts and Techniques in Modern Geography 56, Norwich: Environmental Publications

Martin D, P Longley and G Higgs (1992) The geographical incidence of local government revenues: an intra-urban case study, *Environment and Planning C* **10**; 253–65

Martin D, P Longley and G Higgs (1994a) The use of GIS in the analysis of diverse urban databases, *Computers, Environment and Urban Systems* **18**(1); 55–66

Martin D, M L Senior and H C W L Williams (1994b) On measures of deprivation and the spatial allocation of resources for primary health care, *Environment and Planning A* **26** (12); 1911–29

Marx R W (1990) The TIGER system: yesterday, today and tomorrow. In R W Marx (ed.) *The Census Bureau's TIGER system*, Bethesda MD: ACSM, pp. 89–97

Masser I and M Blakemore (eds) (1991) *Handling geographic information: methodology and potential applications*, Harlow: Longman

Masser I and C Hobson (1992) Britain's Regional Research Laboratories 1992–1995, *Mapping Awareness* **6**; 62–4

Masser I and C Hobson (1993) Where are they now? the diffusion of Regional Research Laboratory staff, *Area* **25**(3); 279–82

Massie K (1993) Solid waste flow modelling: integrating ARC/INFO with Statistical Analysis Software (SAS). Proceedings of Thirteenth Annual ESRI User Conference, vol. 3, Redlands: ESRI, pp. 65–77

Mayo W L (1964) The development of secondary school geography as an independent subject in the United States and Canada. Dissertation Abstracts 25 7027–8, University Microfilms Order No. 65–5349, University Microfilms, Ann Arbor, MI, USA

Mertz G (1993a) Geographic reports can demonstrate compliance, refute redlining charges, *Business Geographics* **1**(4); 18–19

Mertz G (1993b) Insurers recognize geographic variables and need for technology, *Business Geographics* **1**(3); 16

Moore G A (1991) Crossing the chasm. National Assessment Governing Board Pre-publication draft of Geography Assessment Framework for the 1994 National Assessment of Educational Progress

Morphet C (1993) The mapping of small-area census data – a consideration of the effects of enumeration district boundaries, *Environment and Planning A* **25**(9); 1267–77

Morris C (1993) Lifestyle Data. In B Leventhal, C Moy and J Griffin (eds) *An introductory guide to the 1991 Census*, The Market Research Society/NTC Publications, pp. 89–93

Moser C A and W Scott (1961) *British towns: a statistical study of their social and economic differences*, Edinburgh: Oliver & Boyd

National Council for Geographic Education (1993) Draft of national geography standards, The Geography Education Standards Project

National Council for Social Studies, Council of State Social Studies Specialists (1994) National survey of course offerings and testing in social studies, kindergarten – Grade 12

National Geographic Society (1988) Geography: an international gallup survey, The Gallup Organization

National Geographic Society (1994) 1993 Geography Education Division Annual Report to the Education Foundation Board

Newell R G (1993) The why and the how of the long transaction. In D R Green, D Rix and J Cadoux-Hudson (eds) *Geographic information 1994*, London: Taylor & Francis, pp. 237–41

NEXPRI (1989) Geographical information systems for landscape analysis research programme. NEXPRI, University of Utrecht, Holland

Nielsen J B, II G Muller and T Gudmundsson (1993) MIKE SAW 21: a spill modelling system integrated with ARC/INFO. Proceedings of Thirteenth Annual ESRI User Conference, vol. 3, Redlands: ESRI, pp. 295–303

Nordbeck S and B Rystedt (1970) Isarithmic maps and the continuity of reference interval functions, *Geografiska Annaler* **52B**; 92–123

Norman P (1969) Third survey of London life and labor: a new typology of London districts. In M Dogan and S Rokkan (eds) *Quantitative ecological analysis in the social sciences,* Cambridge MA: MIT Press, pp. 371–396

Nyerges T (1992) Coupling GIS and spatial analytical models, Proceedings of the 5th International Spatial Data Handling Symposium, University of South Carolina, USA, pp. 534–43

Oberg S and P Springfeldt (1991) *The population of Sweden*, Stockholm: SNA Publishing

OPCS (1992a) 1991 Census Definitions Great Britain, CEN 91 DEF, HMSO, London

OPCS (1992b) ED/Postcode Directory: Prospectus 1991, Census User Guide 26, OPCS, Fareham

OPCS (1993) Report on review of statistical information on population and housing (1996–2016). Occasional Paper 40, OPCS, London

Openshaw S (1983) Multivariate analysis of census data, the classification of areas. In D Rhind (ed.) *A Census user's handbook*, London: Methuen, pp. 243–64

Openshaw S (1984a) Ecological fallacies and the analysis of areal census data, *Environment and Planning A* **16**; 17–31

Openshaw S (1984b) *The modifiable areal unit problem*. Concepts and Techniques in Modern Geography 38, Norwich: Geo Books

Openshaw S (1989a) Learning to live with errors in spatial databases. In M Goodchild and S Gopal (eds) *Accuracy of spatial databases*, London: Taylor & Francis, pp. 263–76

Openshaw S (1989b) Making geodemographics more sophisticated, *Journal of the Market Research Society* **31**; 111–31

Openshaw S (1991a) Developing appropriate spatial analysis methods for GIS. In D J Maguire, M F Goodchild and D W Rhind (eds) *Geographical information systems: principles and applications*, London: Longman, pp. 389–402

Openshaw S (1991b) A spatial analysis research agenda. In I Masser and M J Blakemore (eds) *Handling geographic information: methodology and potential applications*, London: Longman, pp. 18–37

Openshaw S (1992) A review of the opportunities and problems in applying neuro-computing methods to marketing applications, *Journal of Targeting, Measurement and Analysis for Marketing* **1**; 170–86

Openshaw S (1993a) Future developments. Paper presented to the Market Research Society Census Interest Group, London, 5 November

Openshaw S (1993b) Special classification. In B Leventhal, C Moy and J Griffin (eds) *An introductory guide to the 1991 Census*, Henley: NTC Publications, pp. 69–82

Openshaw S (1994a) Two exploratory space–time–attribute pattern analysers relevant to GIS. In S Fotheringham and P Rogerson (eds), *Spatial analysis and GIS*, London: Taylor & Francis, pp. 83–104

Openshaw S (1994b) Developing smart and intelligent target marketing systems, Part 1 *Journal of Targeting, Measurement and Analysis for Marketing* **2**; 289–90

References

Openshaw S (ed.) (1995) *Census user's handbook*, London: Longman

Openshaw S and A Craft (1991) Using Geographical Analysis Machines to search for evidence of clusters and clustering in childhood leukaemia and non-Hodgkin lymphomas in Britain. In G Draper (ed.) *The geographical epidemiology of childhood leukaemia and non-Hodgkin lymphomas in Great Britain. 1966–83* Studies in Medical and Population, 53, OPCS, HMSO, London, pp. 109–26

Openshaw S, A Cross and M Charlton (1990) Building a prototype geographical correlates exploration machine, *International Journal of Geographical Information Systems* **3**; 297–312

Openshaw S, D Cullingford and A A Gillard (1980) A critique of the national census classifications of OPCS-PRAG, *Town Planning Review* **51**; 421–39

Openshaw S and J G Goddard (1987) Some implications of the commodification of information and the emerging information economy for applied geographical analysis in the UK, *Environment and Planning A* **19**; 1423–40

Openshaw S, F Sforzi and C Wymer (1985) A multivariate classification of individual household census data for Italy, *Papers of the Regional Science Association* **58**; 113–25

Ordnance Survey (1993) Address Point makes first delivery, *Ordnance Survey Data News* Winter 93/94; 4–5

Ottoson L and B Rystedt (1991) National GIS programmes in Sweden. In D J Maguire, M F Goodchild and D W Rhind (eds) *Geographical information systems: principles and applications*, London: Longman, pp. 39–46

OXIRM (1990) The future of Meadowhall Shopping Centre: Further projections. Research Paper D6, Oxford Institute for Retail Management, Templeton College, Oxford

Park R E, E W Burgess and R D McKenzie (eds) (1925) *The city*, Chicago: University of Chicago Press

Parrot R and F P Stutz (1991) Urban GIS applications. In D J Maguire, M F Goodchild and D W Rhind (eds) *Geographical information systems: principles and applications*, London: Longman, pp. 247–60

Peutherer D (1988) Developing small area information systems for service and resource planning in Strathclyde Regional Council. Paper presented to the fifth Regional Science Association International Workshop on Strategic Planning, Enschede, Netherlands

Peucker T K and N Chrisman (1975) Cartographic data structures, *The American Cartographer* **2**(1); 55–69

Polk Direct (1993) Niches. Product brochure, Polk Direct, Detroit

Randall J and P Furness (1993) Geodemographics and the Citizen's Charter. Paper presented to the Market Research Society Census Interest Group, London, 5 November

Raper J F, D W Rhind and J W Shepherd (1992) *Postcodes: the new geography*, London: Longman

Rawling E (1992) The making of a National Geography Curriculum. *Geography*. **77**(4); 292–309

Redfern P (1989) Population registers: some administrative and statistical pros and cons, *Journal of the Royal Statistical Society A* **152**(1); 1–28

Rees P H (1972) Problems of classifying sub-areas within cities. In B J L Berry and K B Smith (eds) *City classification handbook: methods and applications*, London: Wiley, pp. 265–330

Reynolds J (1991) GIS in the commercial and financial world. In J Cadoux-Hudson and D I Heywood (eds) *Geographic information 1991*, Yearbook of the Association for Geographic Information, London: Taylor & Francis, pp. 176–80

Rhind D W (1991) Counting the people: the role of GIS. In D J Maguire, M F Goodchild and D W Rhind (eds) *Geographical information systems: principles and applications*, London: Longman, pp. 127–37

Rhind D W (1992) Data access, charging and copyright and their implications for geographical information systems, *International Journal of Geographical Information Systems* 6(3); 13–30

Ripley B D (1981) *Spatial statistics*, New York: Wiley

Roan N (1993) Mobile telecommunications. In G Castle (ed) *Profiting from a geographic information system*, Fort Collins, USA: GIS World, pp. 177–91

Robbins M (1993) Notes from a personal conversation with N S Bryan

Robson B T (1969) *Urban analysis: a study of city structure with special reference to Sunderland*, Cambridge: Cambridge University Press

Roper C (1993) History divides; geography unites, *GIS Europe* 2(1); 4

Rowlingson B S and P J Diggle (1991) SPLANCS: spatial point pattern analysis code in S-PLUS. North West Regional Research Laboratory, Research Report 22, NWRRL, Lancaster

Rowlingson B S, R Flowerdew and A C Gatrell (1991) Statistical spatial analysis in a geographical information system framework. North West Regional Research Laboratory, Research Report 23, NWRRL, Lancaster

Runnels D (1993) Geographic-based system rescues insurers, *Business Geographics* 1(1); 30–2

Sadler G J and M J Barnsley (1990) Use of population density data to improve classification accuracies in remotely-sensed images of urban areas. Proceedings of the First European Conference on Geographical Information Systems, Utrecht: EGIS Foundation, pp. 968–77

Santoro P J (1993) Using GIS to map, track, and attack the jeans business, GIS in Business '93 Conference Proceedings, pp. 93–6

Sherwood-Bryan N (1993a) What are 'business geographics' anyway? *Business Geographics* 1(1); 24–5

Sherwood-Bryan N (1993b) Putting business GIS users under the microscope, *Business Geographics* 1(5); 8–9

Sherwood-Bryan N (1993c) Radio wave propagation meets business geographics, *Business Geographics* 1(4); 38–9

Sherwood-Bryan N (1993d) Real estate–GIS battle scars fade as bridge building progresses, *Business Geographics* 1(3); 8

Sherwood-Bryan N (1994) Exposed . . . the BG reader, *Business Geographics* 2(4); 10

Shevky E and W Bell (1955) *Social area analysis: theory, illustrative application and computational procedures,* Stanford: Stanford University Press

Shevky E and M Williams (1949) *The social areas of Los Angeles: analysis and typology*, Los Angeles: University of California Press

Sleight P (1992) Classifying people: UK geodemographic systems in 1992, *Admap* May

Sleight P (1993a) Targeting customers: how to use geodemographic and lifestyle data in your business, Henley-on-Thames: NTC Publications

Sleight P (1993b) The new classifications. Paper presented to the Market Research Society Census Interest Group, London, 5 November

Sleight P and B Leventhal (1989) Applications of geodemographics to research and marketing, *Journal of the Market Research Society* **31**; 75–101

Sleight P and B Leventhal (1993) An introduction to multivariate analysis techniques and their application in direct marketing, *Journal of Targeting, Measurement and Analysis for Marketing* **1**(1); 37–53

Statistics Canada (1992) 1991 Geography Catalogue, Statistics Canada, Ottowa

Stover R N (1993) Document management: a pseudo-GIS, *Business Geographics* **1**(5); 24

Strachan A (1994) Teaching Company programme between the Midlands Regional Research Laboratory and Dixons Stores Group Ltd. In D Bond and G Robinson (eds) *The Regional Research Laboratories Network 1994*, Coleraine: University of Ulster

Tavakoli A (1993) GIS eases banking's regulatory compliance efforts, *Business Geographics* **1**(1); 22

Tavakoli H (1993) Mergers and acquisitions soar, increasing need for geographic analysis, *Business Geographics* **1**(2); 26

Theodorsen G A (ed.) (1961) *Studies in human ecology*, New York: Harper & Row

Thrall G I and S E Thrall (1993) Commercial data for the business GIS, *Geo Info Systems* July/August; 63–8

Thrall G I and G Smersh (1994) Business GIS data, Part Four: attribute data, *Geo Info Systems* March; 62–70

Tobler W R (1979) Smooth pycnophylactic interpolation for geographical regions, *Journal of the American Statistical Association* **367**; 519–30

Tomlin C D (1991) Cartographic modelling. In D J Maguire, M F Goodchild and D W Rhind (eds) *Geographical information systems: principles and applications*, London: Longman, pp. 361–74

Tomlinson R F (1987) Current and potential uses of geographical information systems: the North American experience, *International Journal of Geographical Information Systems* **1**(3); 203–18

Tryon R C (1955) *Identification of social areas by cluster analysis*, Berkeley: University of California Press

Tufte E R (1983) *The visual display of quantitative information*, Connecticut: Graphics Press

Unwin D (1981) *Introductory spatial analysis*, London: Methuen

Unwin T (1986) Attitudes towards geographers in the graduate labour market, *Journal of Geography in Higher Education* **10**(2); 149–57

Unwin T (1992) *The place of geography*, Harlow: Longman

Upton G and B Fingleton (1985) *Spatial data analysis by example, Volume 1. Point pattern and quantitative data*, New York: Wiley

US Department of Education National Assessment Governing Board (1993) *Geography Assessment Framework for the 1994 National Assessment of Education Progress (Pre-Publication Draft)*, available from GIS World Inc, Fort Collins, Co, USA

van Arsdol M D, S F Camilleri and C F Schmid (1958) The generality of urban social area indexes, *American Sociological Review* **23**; 277–84

Walker D (ed.) (1993) The Regional Research Laboratory Network 1993, Department of Town and Regional Planning, University of Sheffield

Waltz D L (1990) Massively parallel AI, Proceedings of National Conference on Artificial Intelligence, Boston, August

Watts P (1994) European geodemographics on the up, *GIS Europe* **3**(4) May; 28–30

Webber R J (1977) An introduction to the National Classfication of Wards and Parishes. Planning Research Applications Group Technical Paper 23, Centre for Environmental Studies, London

Webber R J (1978) The National Classifications of Residential Neighbourhoods, An Introduction to the Classification of Wards and Parishes. PRAG Technical Paper TP23, Centre for Environmental Studies, London

Webber R J (1985) The use of census-derived classifications in the marketing of consumer products in the United Kingdom, *Journal of Economic and Social Measurement* **13**; 113–24

Webber R J (1989) Using multiple data sources to build an area classification system: operational problems encountered by MOSAIC, *Journal of the Market Research Society* **31**; 103–9

Webber R J (1992) Streets ahead of the rest, *Precision Marketing* 7 December

Webber R J (1993) Building geodemographic classifications. Paper presented to the Market Research Society Census Interest Group, London, 5 November

Webber R J and J Craig (1978) Socio-economic classifications of local authority areas. Studies in Medical and Population Subjects 35, OPCS, London

Weibul R and M Heller (1991) Digital elevation modelling. In D J Maguire, M F Goodchild and D W Rhind (eds) *Geographical information systems: principles and applications*, London: Longman, pp. 269–97

Wellar B (1993) GIS fundamentals. In G H Castle, III (ed.) *Profiting from a geographic information system*, Fort Collins, US, GIS World Inc., pp. 3–22

Wells W (1972) *Life styles and psychographics*, Chicago: University of Chicago Press

Wilson A G (1974) *Urban and regional models in geography and planning*, London: Pion

Wilson A G (1981) *Catastrophe theory and bifurcation: applications to urban and regional systems*, London: Croom Helm

Wilson A G, J D Coelho, S M MacGill and H C W L Williams (1981) *Optimisation in location and transport analysis*, Chichester: Wiley

Wood S J (1990) Geographic Information System development in Tacoma. In H Scholten and J C H Stillwell (eds) *Geographic information systems for urban and regional planning*, Dordrecht: Kluwer

Worrall L and L Rao (1991) The Telford Urban Policy Information Systems Project. In L Worrall (ed.), *Spatial analysis and spatial policy using geographical information systems*, London: Belhaven pp. 127–51

Wrigley N (1993) Retail concentration and the internationalization of British grocery retailing. In R D F Bromley and C J Thomas (eds) *Retail change: contemporary issues*, London: UCL Press, pp. 41–68

Wrigley N, C Guy, R Dunn and L O'Brien (1985) The Cardiff consumer panel: methodological aspects of the conduct of a long-term survey, *Transactions, Institute of British Geographers* **10**(1); 63–76

Zhu X and R Healey (1992) Towards intelligent spatial decision support: integrating geographic information systems and expert systems. Proceedings of the GIS/LIS '92 Annual Conference, pp. 877–86

SUBJECT INDEX

AUTHOR INDEX